SpringerBriefs in Geography

SpringerBriefs in Geography presents concise summaries of cutting-edge research and practical applications across the fields of physical, environmental and human geography. It publishes compact refereed monographs under the editorial supervision of an international advisory board with the aim to publish 8 to 12 weeks after acceptance. Volumes are compact, 50 to 125 pages, with a clear focus. The series covers a range of content from professional to academic such as: timely reports of state-of-the art analytical techniques, bridges between new research results, snapshots of hot and/or emerging topics, elaborated thesis, literature reviews, and in-depth case studies.

The scope of the series spans the entire field of geography, with a view to significantly advance research. The character of the series is international and multidisciplinary and will include research areas such as: GIS/cartography, remote sensing, geographical education, geospatial analysis, techniques and modeling, landscape/regional and urban planning, economic geography, housing and the built environment, and quantitative geography. Volumes in this series may analyze past, present and/or future trends, as well as their determinants and consequences. Both solicited and unsolicited manuscripts are considered for publication in this series.

SpringerBriefs in Geography will be of interest to a wide range of individuals with interests in physical, environmental and human geography as well as for researchers from allied disciplines.

Long Zhou · Bin Li · Sihong Li · Ngan Leng Lei ·
Kengfong Cheong

Urban and Regional Cooperation and Development

Challenges and Strategies for the Planning
and Development of the Guangdong–Macao
Intensive Cooperation Zone in Hengqin Island

Long Zhou
Faculty of Innovation and Design
City University of Macau
Macau SAR, China

Bin Li
Faculty of Innovation and Design
City University of Macau
Macau SAR, China

Sihong Li
Faculty of Innovation and Design
City University of Macau
Macau SAR, China

Ngan Leng Lei
Institute for Research
of Portuguese-Speaking Countries
City University of Macau
Macau SAR, China

Kengfong Cheong
Faculty of Innovation and Design
City University of Macau
Macau SAR, China

ISSN 2211-4165   ISSN 2211-4173 (electronic)
SpringerBriefs in Geography
ISBN 978-981-19-8060-2   ISBN 978-981-19-8061-9 (eBook)
https://doi.org/10.1007/978-981-19-8061-9

© The Author(s) 2023. This book is an open access publication.

Open Access This book is licensed under the terms of the Creative Commons Attribution 4.0 International License (http://creativecommons.org/licenses/by/4.0/), which permits use, sharing, adaptation, distribution and reproduction in any medium or format, as long as you give appropriate credit to the original author(s) and the source, provide a link to the Creative Commons license and indicate if changes were made.

The images or other third party material in this book are included in the book's Creative Commons license, unless indicated otherwise in a credit line to the material. If material is not included in the book's Creative Commons license and your intended use is not permitted by statutory regulation or exceeds the permitted use, you will need to obtain permission directly from the copyright holder.

The use of general descriptive names, registered names, trademarks, service marks, etc. in this publication does not imply, even in the absence of a specific statement, that such names are exempt from the relevant protective laws and regulations and therefore free for general use.

The publisher, the authors, and the editors are safe to assume that the advice and information in this book are believed to be true and accurate at the date of publication. Neither the publisher nor the authors or the editors give a warranty, expressed or implied, with respect to the material contained herein or for any errors or omissions that may have been made. The publisher remains neutral with regard to jurisdictional claims in published maps and institutional affiliations.

This Springer imprint is published by the registered company Springer Nature Singapore Pte Ltd.
The registered company address is: 152 Beach Road, #21-01/04 Gateway East, Singapore 189721, Singapore

Preface

China's central authorities have recently issued the master plan for constructing the Guangdong–Macao Intensive Cooperation Zone at Hengqin Island in September 2021. As China's first and last European colony and one of China's two special administrative regions (SARs), Macao has developed the gambling industry seven times larger than that of Las Vegas. However, the problem of the homogeneous industrial structure and the urgent need to promote sustainable economic growth by regional cooperation have been important theoretical and practical issues discussed by scholars and policymakers. The Guangdong–Macao Intensive Cooperation Zone will be managed under special customs supervision between two border lines and is expected to diversify Macao's economy. This research firstly introduces the newly unveiled Guangdong–Macao Intensive Cooperation Zone with details as a special mode of the regional collaborative development that is committed to be mutually beneficial to both sides with different political and economic systems. Then, this research dissects the theory of regional synergistic development and its applications in a number of international comparative and cross-interdisciplinary case studies worldwide. Finally, from the perspective of land use, transportation connection and social service, this study thoroughly explores the challenges and strategies to implement the new cooperation model within the framework of the 'One Country, Two Systems', two customs and two currencies to achieve a win-win situation by using updated first-hand data collected by literature review, case study, field survey, spatial analysis and interview.

In the content of this book, Chapter 1 presents the geographic location of Macao and Hengqin, the historical changes in industries and industries, which leads to the critical areas of cooperation in urban infrastructure construction and higher-level planning and presents the challenges of future cooperation between the two regions in terms of policies, funds and talents. Chapter 2 presents the concept and theoretical background of synergistic development of metropolitan areas in the form of city clusters from the perspective of urban planning. Chapter 3 presents the similarities and differences between the planning systems of Macao and Hengqin in the context of 'One Country, Two Systems' and analyses international and domestic cases of cross-border cooperation to offer recommendations for the future planning of the two regions. Chapter 4 introduces the current situation of land use types and industrial distribution in Macao and Hengqin and predicts the development scenarios in future. Chapter 5 analyses the accessibility and convenience of public transportation and service facilities in Macao and Hengqin. Chapter 6 provides a theoretical basis for a more rational allocation of urban green infrastructure resources in the future by quantifying the value of urban ecosystem services and the promotion of complementary alignment of ecosystem services between Macao and Hengqin.

This material can be regarded as a useful handbook for both general readers who want to learn the recent urban development in Guangdong–Hong Kong–Macao Great Bay and graduate students in the academic field of urban and regional studies aiming to understand the theory of regional synergistic development. By studying the special case and situation in the Guangdong–Macao Intensive Cooperation Zone, this research expands the aforementioned theory and offers novel insights into its application and practice. Besides, this research provides policymakers and planners with specific reference, recommendations and experience to develop Hengqin Island.

Macau SAR, China

Long Zhou
Bin Li
Sihong Li
Ngan Leng Lei
Kengfong Cheong

Acknowledgments This research and publication are funded by Macau Higher Education Fund (HSS-CITYU-2021-04). We would like to express our deepest gratitude to Miss Wang Lu and Miss Loi Esme for organising the symposium in June 2021 and all the experts presented in the symposium, who provide us invaluable information and suggestions for this research. We would like to thank Miss Zhu Huiyu and Mr Chen Zeyu for assisting in this research.

Contents

1	**Background of the Guangdong–Macao In-Depth Cooperation Zone in Hengqin**	1
	1.1 Background of the Guangdong–Macao In-Depth Cooperation Zone in Hengqin	1
	1.2 Significance of Hengqin's Development and Its Foundation of Success	8
	1.3 Key Cooperation Areas in the Guangdong–Macao In-Depth Cooperation Zone	9
	1.4 Challenges to the Development of the Hengqin–Macao Cooperation Zone	13
	References	15
2	**Theoretical and Practical Research in the Context of Regional Synergistic Development**	17
	2.1 Introduction to Related Theories	17
	2.2 Cases and Roles	23
	2.3 Hengqin and Macao Synergy: Advantages and Difficulties	29
	References	32
3	**Cooperation Planning System for Hengqin and Macao**	35
	3.1 Introduction on Planning Systems	35
	3.2 Case Studies	36
	3.2.1 Cross-Border Planning Coordination Within the EU	36
	3.2.2 Cross-Border Planning Coordination Between Germany and Poland	37
	3.2.3 U.S.–Mexico Cross-Border Planning Coordination	38
	3.2.4 Transboundary Planning Coordination Between Singapore and Johor, Malaysia	38
	3.2.5 Spatial Coordination Between Guangzhou and Foshan in China	39
	3.3 History and Background of Cooperation Schemes	40

		3.3.1	Development History and Relationship Amongst Guangdong, Macao, Zhuhai and Hengqin	40
		3.3.2	Analysis of the Structural Function of Planning in Hengqin (As a City in Mainland China)	41
		3.3.3	Analysis of the Structural Function of Planning in Macao	42
	3.4	Characteristics of Two Systems and Cooperation Schemes		42
		3.4.1	Legal System of Planning	42
		3.4.2	System of Plan Formulation	43
		3.4.3	System of Planning Management	44
		3.4.4	System of Planning Practice	45
	3.5	Differences Between the Masterplans of Hengqin and Macao		46
	3.6	Strategies of Cooperation Planning Systems for Hengqin and Macao		48
	References			49
4	**Industrial Spatial Synergy Development in Hengqin and Macao**			51
	4.1	Land Use Types in Macao and Hengqin		51
	4.2	Macao and Hengqin Industry and POI Distribution		53
	4.3	Analyses of the Industrial Structures of Macao and Hengqin		53
	4.4	Diversity Analysis of the Urban Block Spatial Function		54
	4.5	Analysis of the Directional Distribution and Central Trend of Residential and Non-Residential Points		56
		4.5.1	Residential Standard Deviational Ellipse with Central Element Analysis	59
		4.5.2	Non-Residential Standard Deviational Ellipses with Central Element Analysis	60
	4.6	Density Analysis of Residential and Non-Residential Points		64
	References			66
5	**Transportation Integration Development in Hengqin and Macao**			67
	5.1	Public Transport Systems in Macao and Hengqin		67
	5.2	Accessibility to Public Transport in Macao and Hengqin		68
	5.3	Accessibility to Public Services in Macao and Hengqin		70
	5.4	Spatial Inequity of Public Service Facilities in Macao and Hengqin		74
	References			85
6	**Ecosystem Services Analysis and Integration in Hengqin and Macao**			87
	6.1	Macao and Hengqin Ecological Services Assessment		87
		6.1.1	Soil Formation	88
		6.1.2	Protecting Species Diversity	89
		6.1.3	Climate Regulation	89
		6.1.4	Environment Purification	90
		6.1.5	Noise Reduction	91
		6.1.6	Climate Regulation	91

	6.1.7 Cultural Service	92
6.2	Analysis of the Value of Terrestrial Ecological Services in Macao and Hengqin	92
6.3	Vision for the Future of Macao and Hengqin	92
References		94

Chapter 1
Background of the Guangdong–Macao In-Depth Cooperation Zone in Hengqin

Abstract Since the handover of Macao, its economy has grown rapidly under the success and advantages of the 'One Country, Two Systems' policy. However, gaming is the single most crucial income source of Macao's economy, and the progress of Macao's economic diversification has been limited by its geographic and demographic limitations. The 'Outline Development Plan for the Guangdong–Hong Kong–Macao Greater Bay Area' is a national strategic development plan that included Macao as one of the key cities to help develop the Greater Bay Area (GBA) and helped Macao break through its development bottleneck. As the 'Hengqin Overall Development Plan' released in 2009, Hengqin has developed rapidly, with increasing population growth which transformed itself from a small town to a developed city with comprehensive infrastructure, high-rise buildings and supporting utilities. Later in 2021, the 'General Plan for Construction of the Guangdong–Macao In-Depth Cooperation Zone in Hengqin' was released to lay out the foundation and goals of utilising Hengqin to supplement Macao in incorporating into the strategic development of the GBA. Both Hengqin and Macao would benefit mutually via in-depth strategic cooperation at a national level. However, this opportunity also comes with various challenges to overcome.

Keywords Guangdong–Macao in-depth cooperation zone · Macao · Hengqin · History

1.1 Background of the Guangdong–Macao In-Depth Cooperation Zone in Hengqin

The plan to develop Hengqin via the cooperation of Guangdong Province and Macao is a strategic decision to help Macao exceed its bottleneck of city development and economic diversification. Historically, Macao has developed much faster than Hengqin in terms of infrastructure, population growth and economy. After years of rapid growth, Macao is seeking new directions to improve the wellbeing of its citizens and create opportunities for future generations. Adjacent to Macao, Hengqin is a desirable place for Macao to extend its future development.

© The Author(s) 2023
L. Zhou et al., *Urban and Regional Cooperation and Development*,
SpringerBriefs in Geography, https://doi.org/10.1007/978-981-19-8061-9_1

The development of the Guangdong–Macao In-Depth Cooperation Zone in Hengqin is part of a more extensive regional development plan for the GBA. In 2019, the General Office of the State Council of the People's Republic of China (PRC) promulgated the Outline Development Plan for the Guangdong–Hong Kong–Macao Greater Bay Area (hereafter called the GBA Outline Plan). The GBA Outline Plan's scope focuses on SARs of Hong Kong and Macao and nine cities (Guangzhou, Shenzhen, Zhuhai, Foshan, Huizhou, Dongguan, Zhongshan, Jiangmen and Zhaoqing) in Guangdong, south of China, commonly known as Pearl River Delta. The GBA has a total land area of more than 56,000 km^2, with a population of more than 86 million, and a combined GDP of more than US$1.9 trillion in 2021 (Statistics of the Guangdong–Hong Kong–Macao Greater Bay Area, 2021). As one of the most economically open and active regions in China, the GBA holds a strategic position in the overall development of the country. The GBA Outline Plan is a strategic proposal to further strengthen the practice of 'One Country, Two Systems' in the region and to incorporate Hong Kong and Macao into the overall development of the country. By leveraging the advantages of Guangdong, Hong Kong and Macao, the wellbeing of citizens in the region could be improved, whilst also maintaining the long-term prosperity and stability of Hong Kong and Macao.

Since the handover of Hong Kong and Macao, the exchange and cooperation of the mainland and the SARs has deepened and continued to strengthen. The GBA has developed rapidly and possesses the fundamentals for developing into a first-class bay area in the world. The GBA is at the forefront of China's economic openness and innovations, and it plays a key role in the One Belt, One Road Initiative. The GBA has one of the most comprehensive transportation systems in the country. With Hong Kong's international shipping centre, world-class ports in Guangzhou and Shenzhen and influential aviation hubs in the area, a highly convenient and modernised transportation network is shaped.

The economic development of the GBA is in a leading position in the country, benefiting from a complete industry chain and economically solid support amongst the cities in the region. With a highly advanced service sector in Hong Kong and Macao and the respective industrial strengths of the Pearl River Delta nine cities, the GBA cities have formed a strategic alliance in innovative industries focusing on modern manufacturing and service as the backbone of the region's development. The GBA is one region that successfully develops trendy and innovative initiatives by means of government support policies and measures. In addition, it has a strong base of research institutes with a solid track record of transforming research findings into industrial use.

Hong Kong is an international finance, shipping, commerce and aviation centre, with a high degree of international exchange and a well-developed commercial network worldwide. Meanwhile, Macao is a world-class tourism and leisure centre continuing to develop as a bridge between China and Lusophone countries. Together with the Pearl Delta nine cities, Hong Kong and Macao are regions with the country's highest degree of freedom and openness to the world and have significant functions and roles in developing the GBA.

The exchange and cooperation amongst Hong Kong, Macao and Pearl River Delta nine cities is steadily increasing. A multi-level and comprehensive cooperation has been established, and remarkable results are achieved in infrastructure development, financial service, innovative technology, education, tourism, environment protection, social services and other areas.

The GBA Outline Plan clearly defines the role of Macao and its strategic development focus in the long term, which is to develop Macao as a world leisure travel centre, a commerce service platform between China and Lusophone countries, to facilitate proper diversification of the economy and create a multi-cultural exchange platform. Within the framework of the GBA Outline Plan, the State Council of the PRC promulgated The General Plan for Construction of the Guangdong–Macao In-Depth Cooperation Zone in Hengqin ('Hengqin General Plan') in 2021, which lists the roles and directions of Hengqin Island to meet the overall goal of the GBA plan. The Hengqin General Plan is a major catalyst to overcome Macao's development limitations, which provides effective measures to ensure Macao's long-term prosperity and success by integrating the development of Macao and Hengqin into the GBA Outline Plan.

Hengqin is an island located in the south of Guangdong Province, which is the largest island amongst Zhuhai's 146 islands. Geographically, Hengqin is adjacent to Macao and Zhuhai and located at the west of Pearl River Estuary, east of Macao, north of Zhuhai free tax zone and connected to Modaomen Channel in the west. Hengqin Island has an area of 106 km^2 (Border Crossing, 2021), an equivalent of 3.6 times of Macao's land area. Hengqin is connected to Macao through Lotus Bridge and connected to Zhuhai through Hengqin Bridge. Until 2020, Hengqin has a population of more than 53,000 people (Hengqin New Territory Seventh National Population Census, 2021). In 2021, the Guangdong–Macao In-Depth Cooperation Zone was established under the jurisdiction of both Macao SAR and Guangdong Provincial Government.

Hengqin Island mainly consists of Dahengqin and Xiaohengqin, which are connected by means of land reclamation. The highest elevation of Hengqin is 457.7 m at Nobui Shan (Zhuhai Hengqin New Territory, 2022), the second tallest mountain in Zhuhai. Spanning Hengqin Island from east to west is Tinmuk River, the largest river in the island, with an average width of 60 m and 80 kms. Hengqin Island is within the subtropical monsoon climate zone, with moderate climate throughout the year. Hengqin Island has a long coastline with plenty of forests. It is an eco-friendly island with various travel attractions and is famous for its oysters' production.

Before Macao opened its port in 1553, residents in Macao made a living by fishing, which shared the same economical patterns of smallholder economy in the nearby regions. Foreign merchant ships would park at Macao to trade every summer, which attracts Chinese merchants, hawkers and porters to the ports. However, trade activities were not active during other times of the year.

The early development of Macao's economy results from its unique geographic location and Ming Dynasty's customs policy and the entrepreneurship spirit of the Portuguese (Fei, 1988). Macao is only a mere a hundred kilometres away from Guangzhou, and vessels could reach Guangzhou and various ports of Guangdong directly. Outgoing vessels could reach other parts of the world directly. In terms of land transportation, Macao's road network could reach the whole Guangdong Province. With its convenience in transportation, Macao had a unique geographical advantage between Mainland China and the West. Besides, Ming's government occasionally restricted mainly Chinese merchants to trade overseas, in which Portuguese merchants filled in the commerce gap between China and the West through Macao. Agricultural crops and handcraft products were exported to the West through this channel, which allowed Macao to be a central hub of the international trade network.

In the late sixteenth century, the Portuguese merchants in Macao established a few major international trade routes, which allowed Macao to be a transfer hub in the international trade network. Through Macao, Portuguese merchants traded goods between China and trading hubs, such as Manila, India, Europe and Japan. Between 1580 and 1590, Portuguese merchants made a good fortune by trading goods, mainly silk and silver (Huang, 1995). By trading in Macao, both Macao and Portuguese merchants have accumulated vast amounts of wealth, which laid a solid foundation for Macao's development. The wellbeing of residents has improved along with decades of prosperity of Macao's economy. At the same time, trade activity also provided Macao the necessary resources and financial support for its urban development and construction, which accelerated its pace to transform itself from a simple port to an international trade hub.

Most of the expenses used in developing Macao's infrastructure, utilities and charity are donated or taxed from Portuguese merchants during that time. The Portuguese built Western-style hospitals, roadways, ports and other infrastructures, which shaped Macao as a Western city in the east. By treating Macao as a platform, the Portuguese dominated maritime trade in East Asia from 1557 to 1641. Macao became the largest portal to trade in China and its economy rapidly developed. The Portuguese made a fortune by using Macao as a transit port in East Asia. However, the Portuguese could not maintain its dominance in the area as various major maritime trade routes were occupied by the Dutch in the seventeenth century. In addition, the Chinese government has restricted Portuguese merchants from trading in Guangdong, which has affected Macao's advantage as a transit port (Yuan, 1988).

In the Ching Dynasty, Macao's foreign trade declined considerably despite Ching Dynasty's leaders offering Macao trade advantages, such as taxation based on vessel size similar to the Ming Dynasty (Yuan, 1988). In 1727, China opened up its ports for foreign trade, which led Macao's foreign trade to a decline. However, the Ching Dynasty's leaders later reverted its policies in 1757, in which only Guangzhou and Macao ports were opened for foreign trades. As a result, Macao's foreign trade was revived (Yuan, 1988). In 1802, the Portuguese government passed a law to offer Portuguese merchants the permit to import opium through Macao. After the First Opium War, Britain was committed to developing Hong Kong to replace Macao as a main transit port in the area, and Macao's advantages in transit vanished.

In 1847, Portugal allowed legal gambling in Macao. Since 1872, the gaming industry has started to become a vital part of Macao's economy, and Macao is well known as the 'Monte Carlo of the East' (Ho, 2011). The financial industry was not well developed in the early days. In 1902, Banco Nacional Ultramarino, S.A. opened the first bank in Macao. In the Republic of China era, the fishery industry became a major industry in Macao's economy. In 1921, as much as 60,000 people relied on fishing to make a living, and the population declined to 20,000 people in 1940 (Yuan, 1988). Part of the fishery products are used for exports and accounted for a significant part of the total exports, which accounted for 26.4% of total exports in 1930. Meanwhile, the three traditional handicraft products, which included joss sticks, matches and firecrackers, also contributed to a large part of Macao's economy, accounting for 37.8% of total exports in 1930. Limited by Macao's geographic nature of shallow harbour, Macao lost its advantages as a transit port after Hong Kong opened its transit port. Although foreign trade activity has declined sharply, Macao's import and export trade, especially food trading with Hong Kong and mainland remained active.

Amongst all the industries, only the gaming industry remained strong. Gaming is prohibited in most of the countries worldwide, including Mainland China. However, gaming remained legal in Macao due to its unique history and position. In the early days of gaming developing, Macao lacked central supervision of the gaming industry. In 1930, Hou Hing Company obtained the franchise to operate casinos in Macao and built the first Western-style casino in the greater area of China (Studies on the Moderate Economic Diversification of Macao, 2020). In 1937, Tai Hing Gaming Company was established and obtained Macao's gaming franchise licence, and it paid 1.8 million patacas of tax every year, which is the main income source of Macao's government.

In the Republic of China period, foreign exports of Macao had totally depraved. The gaming industry became the single most important centre of Macao's economy. After World War II and the establishment of the PRC, people who fled from the mainland during the war started returning to China, and the population of Macao went on decline.

Development of other industries are lagging behind during the prosperous development of the gaming industry. In the 1950s, joss sticks, matches and firecrackers handicraft making were three traditional industries. In 1957, Portugal passed a bill to allow Macao's products to be imported to Portugal area tax free, which attracted plenty of merchants who invested in Macao.

In 1962, the company established by Stanley Ho won the licence to operate the gaming franchise in Macao. According to the contract, his company should invest at least three million patacas and pay a tax of more than 3 million patacas to the government every year from 1962 to 1964. Part of the tax was used for charity, in which most were used for Macao's economic development, including world-class casinos, hotels and transportation infrastructure, to attract people from Hong Kong to visit Macao (Yuan, 1988).

The gaming industry is the main income stream of the Macao government. The gaming tax increased progressively from 6.9 million patacas in 1975 to 71 million patacas in 1980, and it contributed to more than 25% of the government's total tax income (Hong Kong & Macao Economy, 1986).

Early modern industry of Macao is the textile industry, in which textile exports contributed to 19% of exports in 1958 and increased to 71.6% in 1969, fishery industry dropped to 9.2%, and the three traditional handicraft (joss sticks, matches and firecrackers) industries only contributed 7% (Hong Kong & Macao, 1986).

The economy of Macao picked up the pace in the 1970s. In 1971, some developed industrial countries offered exemptions to various developing countries. Macao was listed as one of the beneficial regions, which led to a blossom of Macao industrial development. In 1976, the European market was Macao's largest export market, accounting for 64.1% of Macao's industrial export. As a result of international trade protectionism, Hong Kong exports were subjected to various restrictions worldwide, and hence, many Hong Kong investors opened factories in Macao to avoid such restrictions. There were 949 modernised factories in 1972 and increased to 1037 factories in 1974. Macao industrial production had reached into a new stage (Yuan, 1988).

In the 1980s, Macao economy grew at a rapid pace. Textile, electronics, toys and artificial flowers account for 91% of export products. The increase in exports benefited from Macao's free market policy and the low-income tax of only 15%, which is 2.5% lower than that of Hong Kong. In addition, the labour cost of Macao was 20–35% lower than that of Hong Kong. Lower living and production costs helped Macao to stay competitive in the international market. In the 1990s, the growth of industries slowed down. Most products exported to western Europe and North America are textile products.

Macao tourism also stimulates the gaming industry and vice versa. The number of tourists increased from 4 million in 1982 to 4.5 million in 1985, and tourism profits were second to the gaming industry. After the handover of Macao in 1999 and Macao's historic region listed in the UNESCO World Heritage Site in 2005, Macao has attracted tourists from around the world, especially Mainland China. Most of the tourists are from Mainland China, Hong Kong and Taiwan (Xinhua Net, 2016).

After the Macao handover in 1999, Macao fully utilised its unique position to facilitate the Chinese–Portuguese commerce exchange and service platform, to deepen the trade cooperation between China and Portuguese-speaking countries. In 2004, Macao and China had a mutual Closer Economic Partnership Arrangement, which allowed products produced in Macao to be imported into China without tax. With different factors and policies that are beneficial to Macao's development, Macao's economy developed rapidly. However, gaming remains the dominant industry in Macao which supports Macao social and economic development.

According to the development plan proposed by the Macao government, the 'Five-Year Plan for Economic and Social Development (2021–2025)' proposed to build Macao as 'one centre, one platform', namely a World Centre of Tourism and Leisure and a Commercial and Trade Co-operation Service Platform between China and Portuguese-speaking countries. The government is eager to develop Macao as a portal to bring tourists from around the world and helps Macao corporations to grow and extend its business outside Macao.

Hengqin Island is the largest island of Zhuhai adjacent to Macao and has long been a key focus development area. In the 1990s, Hengqin Island was included into the economic zone of Zhuhai, and the Zhuhai Economic Development Zone was established. In 1992, Guangdong government confirmed it would develop Hengqin as a critical development area. After the handover of Macao, Hengqin Island became the bridgehead of the cooperation between Guangdong and Macao.

The Hengqin development area started in 1992. Originally, the area of Hengqin was only 48 km^2. Amongst the current 106 km^2 of land, more than 60% are manmade in the 1980s and 1990s by means of land reclamation. As one of the islands of Zhuhai, there were very few residents in Hengqin. Most residents made a living by fishing, agriculture and small-scale commerce. There was rarely any industrial and city construction in Hengqin at that time.

In 1992, Guangdgong Provincial Government released the 'proposal to enhance openness'. According to the proposal, Hengqin, western Zhuhai, Guangzhou Nansha, and Huizhou Daya Bay were selected as the four key open areas. The development of Hengqin was brought to discussion, and the 'Zhuhai Hengqin Economic Development Management Council' was established as a means of attracting foreign investment and development. By 2003, there were 908 enterprises, with approximately 1.3 billion RMB of capital invested in Hengqin. At the same time, telecommunications and power supply network infrastructure were built to meet the needs of the businesses in Hengqin. Infrastructures, such as bridges, roadways and border gates, were also simultaneously built. Foreign investors were attracted to start business in Hengqin as infrastructure improved. Various resorts and parks were built subsequently by foreign investors to attract visitors. In 2004, Guangdong and Macao leaders had discussed developing Hengqin as an international, diverse, open leisure area and as an extension to Macao's leading industry. However, the direction was changed back and forth before any of the plans had ever come into reality. In 2005, Guangdong designated Hengqin as 'Pearl River Delta Economic Hengqin Cooperation Zone', utilising Hengqin as a platform for development. The position of Hengqin was to cooperate with Hong Kong and Macao, servicing Zhuhai, and act as an example of an innovative region to the mainland. At the same time, the 'Outline Plan of Hengqin Development' was finished. In 2006, the Guangdong government passed the 'Outline Plan of Hengqin Development'. After a few years in 2008, the development of Hengqin was still in its early stage. Most early investments focused on travel and leisure projects. The '11th Five-Year Plan' stated that Hengqin should focus on developing modern service and high-end manufacturing sectors. In 2009, the 'Hengqin Overall Plan' stated Hengqin's development goal, to systematically and scientifically develop Hengqin's industry, which should be in harmony with the environment. By developing characteristic and compact development, the functions of subareas of Hengqin could be completely utilised (Liu & Un, 2010). In 2012, the first amusement park in Hengqin, the Chimelong Ocean Kingdom opened its doors to the public. Located in the south of Hengqin Island, it has a footprint of more than 3 million square metres. It includes a convention centre, an aquarium, hotels and an amusement park. In the same year, the first deal of land bided by Macao's corporation was completed. The property included a shopping mall, restaurants and leisure and entertainment facilities, with a footprint of 150,000 m^2, covering an investment capital of 1600 million RMB. In 2021, the Hengqin General Plan clearly stated the four new industries for Hengqin development, which focused on growing scientific and technological R&D and high-end manufacturing industries, such as traditional Chinese medicine, the culture, tourism, convention and exhibition industry, commerce and trade, and modern financial sectors.

1.2 Significance of Hengqin's Development and Its Foundation of Success

The decision to develop Hengqin is a critical step of the GBA Outline Plan, which helps Macao to incorporate into the strategic overall plan at a national level. The development of the Zhuhai–Macao Cooperation Zone is an innovative and ambitious step which is of great importance to the future of Macao and Zhuhai. Both Hengqin and Macao have solid foundations and strengths in various aspects to ensure the success of the Hengqin Cooperation Zone.

Macao has a land area of only 33 m^2 and is the smallest region amongst the GBA's cities. With a population density of 19,737 people per square kilometres of land area (World Bank Data, 2021), Macao is one of the most densely populated regions in the world. The scarce land has greatly limited the room of Macao's development and the wellbeing of residents. In 2009, the General Office of the State Council of the PRC approved Macao's proposed plan to perform land reclamation of 3.6 km^2 of land (Macao Government Information Bureau, 2009), which accounts for approximately 10% of land area in Macao to help relieve Macao's land shortage problem.

Hengqin has a land area of 106 km^2, which is only about one-ninth of Zhuhai's area, but it is already three times the area of Macao. For years, Macao land development has faced the challenges of land shortage, which has led to the diverse and mixed-use nature of existing land development in Macao. To improve the wellbeing of Macao residents, more land for development is urgently needed. By developing the Guangdong–Macao In-Depth Cooperation Zone, Macao could utilise the vast underdeveloped land of Hengqin to relieve its stress on land demand. By easing the customs clearance procedures, creating more room and eco-friendly environment, Hengqin could attract more Macao residents to relocate to Hengqin for bigger, better living conditions and quality of life. With more population, Hengqin could also benefit from more human and business activity at the same time.

Economically, Macao has a long history and rich experience in the gaming industry. Since the handover of Macao, the economy has undergone an explosive growth period in which the gaming revenue increased from MOP 55 billion in 2001 to MOP 446 billion in 2019. During the same period, the GDP per capita increased from MOP 127,015 to MOP 350,445, respectively (Macao Statistics & Census Service, 2022), which ranks as one of the highest GDP per capita worldwide. In the past 20 years, the development and growth of the gaming industry in Macao is unprecedented in any part of the world. However, other industries, even those affiliated with gaming, such as hospitality, catering and convention, could not keep up with the rapid growth of the gaming industry. In terms of economic contribution, the gaming industry became a major income source of Macao's economy, whilst GDP contribution of other industries are shrinking. Since 2001, the Macao government has promoted economic diversification by means of financial and policy support (Hou, 2006). In contrast to the growth of gaming development, the progress of economic diversification is not satisfactory despite tremendous efforts by the Macao government. The two key factors that hinder Macao's diversifications are the lack of land and human resources. By means of the in-depth cooperation between Guangdong Province and Macao, Hengqin is fertile ground to supplement the weakness of Macao to cultivate and diversify its industries.

Hengqin, mainland and Macao could supplement each other and form a strong alliance to form a mutual relationship with competitive edge in developing innovative industries. With the abundance of land Hengqin has, labour resources of Mainland China, capital and talents from Macao, Hengqin and Macao could supplement each other via in-depth cooperation.

Different from Pudong New District in Shanghai and Binhai New District in Tianjin, which are solely running on a 'one country system'. Meanwhile, the Hengqin Cooperation Zone has the advantages of being adjacent to Hong Kong and Macao, which are running in 'One Country, Two Systems' mode, and serve as connection hubs to international countries. Hence, Hengqin could enjoy the benefits of accessing the resources of Guangdong Province and Mainland China and the strong international ties of Hong Kong and Macao with other countries. The measures and policies used in the cooperation zone are one of its kind in the history of the nation and have a demonstration effect in terms of increased openness, innovation and mixed management in other regions in the country.

1.3 Key Cooperation Areas in the Guangdong–Macao In-Depth Cooperation Zone

With the release of the Hengqin Development Overall Plan, The Outline Development Plan for the Guangdong–Hong Kong–Macao GBA and the Guangdong–Macao In-Depth Cooperation Zone in Hengqin Overall Plan, Hengqin has undergone a drastic change and rapid development in the past decade. Hengqin is developed from a small town with thousands of populations to a developed city, supported by world-class infrastructure and facilities. In addition, various innovative and breakthrough policies were also imposed to provide further convenience and advantages to Macao residents to live and work in Hengqin, further attracting foreign capital to do business in Hengqin. With the improvement in infrastructure and policy complement, the willingness to work and reside in Hengqin increased over years.

The Hong Kong–Zhuhai–Macao Bridge (HZMB) is a 55 km bridge-tunnel connecting Hong Kong, Macao and Zhuhai, which reduced the travel time from Hong Kong to Macao from three hours to half an hour, and has allowed these three regions to stay within a one-hour living circle (Gabinete para o Desenvolvimento de

Infraestruturas, 2017). HZMB has connected strategic development cities in Mainland China, Hong Kong and Macao. Given the convenience of transportation in the region, more international travellers are expected to visit Hengqin and Macao and become world-class travel destinations. By improving transportation infrastructure, more cooperation and exchange amongst Mainland China, Hong Kong and Macao are expected.

The customs clearance policy is a key part of Hengqin cooperation zone development. Easing of customs clearance will greatly encourage the cooperation and exchange of Macao citizens between Hengqin and Macao. By exploring the gradual easing of customs clearance of people and goods between Hengqin and Macao, the merging of the Hengqin Cooperation Zone will move at a faster pace than ever. In 2021, a flyover between University of Macau (UM)-Hengqin campus and Hengqin port area is proposed and currently under construction, aiming to provide another connection gateway to Hengqin from Macao (Government Information Bureau of the Macao SAR, 2022). In Phase 2 of the Hengqin–Macao port expansion, the number of car lanes will greatly increase to 30 lands, which would greatly reduce the car clearance time to approximately one minute per car and provide great convenience to drivers commuting between Hengqin and Macao (Exmoo News, 2022).

In addition, the light rail transit extension of Macao and Hengqin is under construction to connect the LRT Taipa line to Hengqin. In the future, the Hengqin line will connect with the Zhuhai light rail and the high-speed railway network connecting the whole country.

The construction of the UM campus in Hengqin gained the approval of the central government in 2009 and started operation in 2013. The UM Hengqin campus was a new attempt of governance in the Hengqin area, in which part of Hengqin is rented to Macao for the construction and operation of UM under Macao's jurisdiction. The UM area is separated from Hengqin's justification by port walls and connected to Macao through an underground tunnel. The location of UM is next to Huandao East Road, across from Hengqin port gate. It has an area of 1.09 kms, a total building area of 820,000 m^2.

The UM-Hengqin campus accommodates more than 12,000 local and international students (University of Macau Registry, 2022). The university is currently a Top 250 university in the world. It is currently home to three state key laboratories: Laboratory of Analog and Mixed-Signal VLSI, Laboratory of Quality Research in Chinese Medicine and Laboratory in the Internet of Things for Smart City. It is one of the fastest growing young universities in Asia, which ranks 26th in young university ranking (Times Higher Education, 2022).

Not only has the university campus' construction helped Macao's higher education development, it has also given significant meaning and long-term effects and even helped deepen the higher education cooperation amongst Mainland China, Hong Kong and Macao. As Macao has lacked the talents and land for higher education development, the campus is a big step forward in Macao's higher education, the construction of the university is a breakthrough in Hengqin–Macao cooperation and an innovative move which accentuates the strengths of the 'One Country, Two Systems' policy. In summary, the university has strong guidance in Macao's diversification and expansion and a significant milestone in Hengqin–Macao cooperation.

The Macao New Neighbourhood Project is a residential neighbourhood project located in the east of Hengqin's Cooperation Zone Central Avenue. It is developed by the Macau Urban Renewal Limited, a company fully owned by the Macao government. The project would provide 4000 residential units with a land area of 190,000 km^2, with a total floor area of 620,000 km^2. The whole project plan comes with retail, education, transportation, medical facilities and community service hubs and is expected to be completed by 2023 (Macau Urban Renewal Limited, 2022). The Overall Plan stated to support Zhuhai and Macao cooperation in building a comprehensive community to support senior healthcare, residential, education, medical service, in which Macao's medical and social coverage system directly could be applied. The project aims to provide the necessary support and convenience for Macao citizens to learn, work, do businesses and live in the GBA. The project consists of 27 buildings, 20–26 storeys high and will be constructed in environmentally friendly methods, such as precasting technology. Approximately 80% of all the units are two-bedroom apartments with areas of approximately 90 m^2. The three-bedroom apartments account for 20% of the units, with areas from 100 to 120 m^2 (Macau Urban Renewal Limited, 2022). The project is six minutes away from the port gate. For the ease of transportation, the project includes approximately 4000 car parks, and residents are qualified to apply for the driving permit to drive in and out of Hengqin. In addition, shuttle buses service will be provided when residents move in in the future. The Macao New Neighbourhood Project is a pioneer project to help Macao citizens to improve living conditions and wellbeing, via the cooperation between Guangdong Province and Macao government. Macao citizens could live, work and enjoy the facilities of the Macao New Neighbourhood, whilst retaining their lifestyle, as the Macao's social and medical coverage will be applied in the neighbourhood for the ease of Macao's citizens.

Macao and Hengqin have their own social policies. The General Plan gives the local government the authoritative power to develop its own policies, including administration, society management and economic management, to build rules and policies that comply with the country's law. The goal is to build policies and rules that are advantageous and suitable for the development of Zhuhai and Macao. Zhuhai is one of the earliest economic zones in China and has its regional law establishment rights. Through 30 years of open development, Zhuhai has established its market economy and system. However, much of its governance is affected by past traditions and local practice. Meanwhile, Macao was under Portuguese governance. Under the 'One Country, Two Systems' establishment after the Handover, Macao kept its own social, economic and law system, with a more comprehensive market economy. Not only is Macao a free-trade zone, with tax exemptions and low transaction cost, it also has a long history of international trade and cultural exchange with other countries. By means of the treaty relationships with other countries, Macao could help Zhuhai and Hengqin to enhance trade relationships with other countries. Although

Macao is a city with small trade volumes, its friendly relationships with other countries, especially European countries are essential to the regional development of the Guangdong–Macao In-Depth Cooperation Zone.

The General Plan for Construction of the Guangdong–Macao In-Depth Cooperation Zone in Hengqin proposed a new governance structure of the cooperation zone. The Guangdong Province and Macao government would each send representatives to form a management committee. Under the committee, a cooperation zone execution committee is established. Guangdong Province would also establish organisations to help manage and maintain order of the Area. The Hengqin–Macao Port is a key infrastructure construction in the cooperation and exchange between Hengqin and Macao. Hengqin will be implementing an innovative customs clearance policy that is unprecedented in China. The keys of this policy include the following: (1) the port between Macao and Hengqin is defined as 'first line', and the port between Hengqin Bridge and other parts of Mainland China is defined as 'second line'; (2) the port control will implement 'co-location arrangement' policy; and (3) provide convenience to Macao citizens travelling between Hengqin and Macao.

The General Plan proposed the idea of 'manage by zones in innovative modes' to build the Hengqin port. The port between Hengqin and Macao will be classified as 'first-line control' where port inspections and clearance will be eased, whilst the ports between Hengqin and mainland are classified as 'second-line control', which is mainly for the customs clearance for customs clearance of exports, inspections, taxation and other monitoring functions. The function of exports clearance and inspections are moved from the 'first line' to the 'second line'. The new customs clearance greatly increased the convenience of people going in and out of Hengqin (State Council of the People's Republic of China, 2021).

The financial market of the cooperation zone will be highly opened to increase the convenience of capital flow and attract more investment in the cooperation zone. The capital is free to import, export and exchange for investment in the area. In addition, new external debt management policies in the area will be explored. Policies will be established to improve the convenience of investment entering the market of the cooperation zone. In addition, the cooperation zone will be a testing point to establish cross-port mutual exchange of Internet data.

The cooperation zone also allowed professionals in finance, architecture, city planning and design, etc. from Macao and other countries to work in Hengqin. The experience and qualifications from outside Mainland China are also recognised. The committee would also promote Macao entrepreneurs to enjoy the same privilege and supporting policies from both Guangdong province and Macao, to invest and start business in the cooperation zone.

For qualified firms stationed in the cooperation zone, a business tax reduction of 15% is offered, which would be beneficial to industries contributing to diversify Macao's economy. For the four key industries in the cooperation zone (tourism, modern service, high-technology enterprise and Chinese medicine), business income tax is exempted from the income earned outside the cooperation zone. For products without imported goods or with import goods lower than 30% of composition, the import tax could be exempted and imported to Mainland China through the 'second-line' port.

For foreign high-end talents or talents on-demand working in the cooperation zone, personal income tax exceeding 15% could be exempted. In addition, personal income tax exceeding Macao's tax would be exempted for Macao citizens working in the cooperation zone.

1.4 Challenges to the Development of the Hengqin–Macao Cooperation Zone

The Overall Plan has positioned Hengqin Cooperation Zone to be a centre of four new innovative industries, which included high-end manufacturing, Chinese medicine, travel and modern finance industries. Although Hengqin and Macao have their strengths in these areas, the difference between Hengqin and Macao in various aspects also poses several challenges. Most of these challenges resulted from different historical backgrounds Hengqin and Macao have. The Guangdong–Macao in-depth cooperation is a breakthrough proposal driven by the Central government to help overcome the challenges in various aspects to ensure the continued prosperity of Macao in the future.

Given the different policies of Mainland China and Macao, many of the developments are hindered. Gaming is the main industry in Macao, but it is prohibited in Mainland China. The difference in policies has affected the development of Guangdong–Macao cooperation. In addition, Macao economy is based on capitalism, whereas the mainland economy is based on communism. From the objective perspective, the difference in systems has prohibited flow and exchange of capital from both places. To overcome these challenges, both places would need to pay a heavy cost on politics, social and economy. In addition, the citizens from the two cities have lived and grown up with very different cultures, traditions and values. As the objectives and perspectives are very different, the laws, city planning and law making are conducted based on their own perspectives and local practice. Not only this has created a single minded situation, Zhuhai citizens would have a mentality of supporting Macao instead of benefiting from this partnership.

Throughout the development history of Hengqin, it has undergone various changes in development overall plan and management authority. Before the release of The General Plan for Construction of the Guangdong–Macao In-Depth Cooperation Zone in Hengqin in 2021, Hengqin was mainly developed by the Zhuhai government. After the General Plan was released, the cooperation zone was jointly governed and managed by the Guangdong Province and Macao government. Over the years, Hengqin development has undergone various changes in development directions and the change in management authorities. In order for the cooperation zone to develop prosperously and successfully, Hengqin development would require a consistent long-term plan, with in-depth cooperation between the Guangdong Province and Macao government.

Mainland has a strict capital control policy, which restricts the flow of capital in and out of the country, whilst Macao allows free flow of capital. The difference in policies create inconvenience for investors and business owners to do business in the cooperation zone. Openness is critical for the development of the modern finance industry, which requires a high degree of freedom in capital movement. On the basis of the successful experience of world's largest financial markets, such as New York, London and Tokyo, etc., the free flow of capital is a key element in a mature and developed financial market to help develop the modern finance industry in Hengqin, the government bodies of Guangdong Province and Macao should cooperate to allow a higher degree of financial freedom in the cooperation zone and provide more incentives for investors and companies to do businesses in Hengqin.

Differences in taxation between mainland and Macao are also barriers to developing the cooperation zone. In general, the mainland has higher corporate and personal income taxes than Macao. To attract business and individuals, Hengqin's Management Committee announced that tax incentives will be provided to firms and individuals doing business and working in Hengqin. Although the incentives greatly decreased the corporate and individual income taxes in Hengqin, the tax is still higher than that of Macao. The difference in tax may lower the interests of Macao's cooperates and individuals to relocate to Hengqin.

To facilitate the development of the cooperation zone, the financial market and capital control policy of Hengqin should be aligned between Hengqin and Macao, to provide more freedom and convenient policies to encourage entrepreneurship and stimulate business activity.

Difference in regulations between mainland and Macao may also slow the development of the cooperation zone. The regulations of industry, such as Chinese medicine, are vastly different in mainland regulations compared with Macao regulations. For example, the new Macao's 'Chinese Medicine Registration Law' stated that the approval and registration of Chinese medicine in Macao recognises Chinese medicine that are approved based on other countries' standards. Meanwhile, Mainland China imposes strict regulations on the safety and quality of Chinese medicine, such as requiring extensive toxicology and clinical trials, to meet the local standards of Chinese medicine approval and registration. To help support industry development in the cooperation zone, the difference in regulations across the four key developing industries should be minimised.

With a relatively small population, both Hengqin and Macao development are limited by lack of human resources and talents. For industries, such as technology and high-end manufacturing, high-skilled labour and professional expertise are amongst the key elements of success. Considering the goal of developing a high-end manufacturing industry, the cooperation zone would require a large number of skilled workers. As Hengqin and Macao are not surrounded with cities with strong high-end manufacturing industries and skilled workers, there is a shortage of skilled workers and talents around the area. Although Hengqin and Macao both have incentives to attract talents and skilled workers to work in Hengqin, the progress of talent attraction to Hengqin is unsatisfactory. On top of financial subsidies, other incentives, such as job opportunities, healthcare, family support, and other factors, should be considered to attract talents to the cooperation zone.

With the release of the Greater Bay Area Overall Plan and the General Plan for Hengqin Cooperation Zone, the cooperation of Guangdong Province and Macao to develop Hengqin will provide valuable opportunities to transform the area. For Macao, the development of Hengqin is critical to its transformation from an economy highly relying on the gaming industry to a diverse economy with multiple high-value modern industries. Although Hengqin has come a long way in terms of its infrastructure and policy development, plenty of challenges still need to be addressed in developing the cooperation zone into a desirable area for Macao residents to reside and work. The continued prosperity of Guangdong Province and Macao would require a highly collaborative and strategic management from both parties.

References

Border Crossing. (2021, September 14). *Guangdong-Macao in-depth cooperation zone in Hengqin.* Retrieved June 30, 2022, from http://www.hengqin.gov.cn/macao_en/tzhzq/zszy/content/post_2996902.html

Exmoo News. (2022, June 07). *Hengqin car lanes expected to extend to 30 lanes.* Retrieved June 30, 2022, from https://www.exmoo.com/article/200240.html

Fei, C. (1988). *Macao 400 years.* Publishing House of Shanghai Academy of Social Sciences.

Gabinete para o Desenvolvimento de Infraestruturas. (2017, December 27). *HongKong-Zhuhai-Macao bridge background information.* Retrieved June 30, 2022, from https://web.archive.org/web/20171227180255/http:/www.gdi.gov.mo/special_project_background.php?id=12

Government Information Bureau of the Macao SAR. (2022, February 14). *Special notice on University of Macao to Hengqin Border Gate Connection.* Retrieved June 30, 2022, from https://www.gcs.gov.mo/detail/zh-hant/N22BNhgTis?1

Hengqin New Territory Seventh National Population Census. (2021, September 17). *Statistics bureau of Guangdong-Macao In-depth cooperation zone in Hengqin.* Retrieved June 30, 2022, from http://www.hengqin.gov.cn/stats/tjfw/zlcx/content/post_3026665.html

Ho, V. (2011). *Macao: Cultural connotations beyond the casino.* (p. 96). City University of HK Press.

Hong Kong and Macao. (1986). In Q. X.-l. Peng. The Commercial Press.

Hou, L. K. (2006). Macao"s industry structure improvement and diversification. *Administration, 19*(72).

Huang, Q. (1995). *Macao history.* Macao Foundation.

Liu, Z. Y., & Un, K. C. (2010). Special studies on Zhuhai-Macao Cooperation to Develop Hengqin. *Macao Economic Association.*

Macao Government Information Bureau. (2009, November 30). Retrieved June 30, 2022, from https://www.gov.mo/zh-hant/news/72027/

Macao Statistics and Census Service. (2022, 5 27). *Gross domestic product (GDP) and per-capita GDP.* Retrieved June 30, 2022, from https://www.dsec.gov.mo/ts/#!/step2/PredefinedReport/en-US/32

Macau Urban Renewal Limited. (2022). *Macau new neighbourhood.* Retrieved June 30, 2022, from https://www.mur.com.mo/en/project/macau_new_neighborhood

The State Council of the People's Republic of China. (2021, September 05). *Release of "The general plan for construction of the Guangdong-Macao in-depth cooperation zone in Hengqin"*. Retrieved June 30, 2022, from http://www.gov.cn/zhengce/2021-09/05/content_5635547.htm

The World Bank Data. (2021). *Population density (people per sq. km of land area)—Macao SAR, China*. Retrieved June 30, 2022, from https://data.worldbank.org/indicator/EN.POP.DNST?locations=MO

Times Higher Education. (2022). *World university rankings*. Retrieved June 30, 2022, from https://www.timeshighereducation.com/world-university-rankings/university-macau

University of Macau Registry. (2022, March 25). *Student figures*. Retrieved June 30, 2022, from https://reg.um.edu.mo/about-reg/facts-and-figures/students-figures/

Xinhua Net. (2016, 01 23). *From Macao arriving passengers declined by 2.6% in 2015*. Retrieved June 30, 2022, from http://www.xinhuanet.com/travel/2016-01/23/c_128659484.htm

Yuan, B. Y. (1988). *A brief history of Macao*. China's Exchange Publishers.

Zhuhai Hengqin New Territory. (2022, June 02). *China (Guangdong) pilot free trade zone*. Retrieved June 30, 2022, from http://ftz.gd.gov.cn/qyjj/content/post_917246.html#zhuyao

Open Access This chapter is licensed under the terms of the Creative Commons Attribution 4.0 International License (http://creativecommons.org/licenses/by/4.0/), which permits use, sharing, adaptation, distribution and reproduction in any medium or format, as long as you give appropriate credit to the original author(s) and the source, provide a link to the Creative Commons license and indicate if changes were made.

The images or other third party material in this chapter are included in the chapter's Creative Commons license, unless indicated otherwise in a credit line to the material. If material is not included in the chapter's Creative Commons license and your intended use is not permitted by statutory regulation or exceeds the permitted use, you will need to obtain permission directly from the copyright holder.

Chapter 2
Theoretical and Practical Research in the Context of Regional Synergistic Development

Abstract This chapter provides a research overview of the related theoretical concepts and practical examples of synergistic regional development. The solution to regional development imbalances from the path of regional cooperation is one of the main objectives of this theory. It provides theories on how regions can develop synergistic collaboration in the areas of economic, environmental and social structures and how these theories can be practised into practical strategic implementation programmes for sustainable regional development. Firstly, we introduce the conceptual and theoretical background of synergistic regional development in the form of urban agglomeration discussed from the perspective of the urban planning discipline. Then we analyse and evaluate cases of synergistic regional development that reflect the research results of regional cooperation. Finally, we explain the advantages and difficulties of forming a synergistic development of Hengqin and Macao.

Keywords Regional synergistic development · Theoretical and practical research · Hengqin and Macao

2.1 Introduction to Related Theories

In urbanisation, a region or city has a certain degree of unbalanced development due to various factors, such as natural conditions, resources, production, innovation capacity and ethnic, religious and cultural differences. It includes the imbalance in the degree of economic development, the level of public services, the innovation potential and the sustainability of development (Fan, 2011), which triggers the need for positive value development. It is called the justice of urban areas. However, synergistic regional development is a homogeneous demand for city–regional justice (Gao, 2016), and synergistic regional development is a path to solving the problem of regional development imbalance. In this review, we aim to explore the theory of synergistic regional development in the context of urban planning discipline and based on the relationship between cities and regions. The following section reviews the theoretical development and concepts of cities and regions for synergistic regional development.

Cities are the products of human civilisation at different stages of development, and the origin of cities needs to be discussed with the development of human civilisation. The origins of the Western city are explored in the book 'The City in History: A Powerfully Incisive and Influential Look at the Development of the Urban Form Through the Ages' by American urban theorist Lewis Mumford. The author argues that there are many manifestations of the process of city production, its functions and the purposes for which it was intended, and it is not easy to summarise it into a single definition. The development of the city has different and rich stages, such as the social core of its embryonic period, the complex forms of its maturation and the decomposition and collapse of its ageing period. Mumford mentions that amongst the early human non-permanent settlement forms, it has two forms related to gods and rituals. In the stage of non-permanent settlement, Mumford believes that before the city became a permanent human settlement, its initial function was to provide meeting places for humans. These locations would periodically bring humans together for sacred events, where humans could interact and communicate. This setting was one of the essential criteria of a city and an evidence of its vitality. Thus, the earliest human ceremonial gathering places, such as sites of ritual and religious veneration. They were the embryonic form of urban development. With development, non-permanent settlement forms transitioned to permanent settlement forms. During the Palaeolithic period, the remains of buildings that seemed to be primitive settlements were found, and the Neolithic culture period was a time when agricultural villages and towns were not yet widely formed; however, by this time, humans knew how to probably choose favourable locations for these future villages and towns, such as locations with abundant water resources, rich fish and shellfish resources and convenient transportation. In the course of later studies, it was discovered that a large number of hills with shells were found in these locations. These form the evidence of permanent settlement, a primitive form of the small village. Then later, during the Neolithic, new village settlement centres emerged, and the primitive villages and surrounding plots constituted new types of settlements. During the Neolithic period, humans began to develop agriculture, invent various technologies and build primitive village structures, including houses, stoves and barns, etc. With these structures and the way of life of people living close together in villages, the order and stability of towns would continue and be transmitted to later cities. All of these prove that the primitive forms of cities already appeared in the primitive Neolithic villages and that cities evolved from these primitive villages. The city began to take shape gradually, and it became richer and more prosperous. The change in the original settlement form proves that early cities existed. At the same time, human beings carried out activities to transform the natural environment. The above factors and evolutionary processes provided a solid foundation for the formation of later cities. The method of human civilisation was an important driving force in the development and formation of cities, which later developed forms of urban organisation thanks to the foundations left by ancestral humans.

The city emerged as a new thing amongst the Palaeolithic and Neolithic communities, which gradually took shape as it developed. As a result of the limitations of the form and ability of the original village, the promotion of an actual city required a new factor to drive it. Innovations were introduced to the primitive villages in the process of development. Under the influence of new factors, the basic form of the primitive village began to become complex and unstable, and the human organisation began to become more complex, and the transition from primitive villages to cities began during this period, similar to the case of the Sumerian states that established slaveholding city-states in the twenty-fourth century A.D. Owing to the rise of cities, which brought together many social functions within a limited geographical environment, it became regular and organised. The composition of the social structure began to be further differentiated, and each part became the prototype of the various constituent structures of urban culture. This was followed by the emergence of castles, such as Khorsabad, an ancient city in northern Iraq, which was a capital city in the seventh century BCE. Finally, in 2500 B.C., all the basic features of the city were formed (Lewis, 1961). There have also been many discussions amongst scholars about the origin of Chinese cities. Although China's urban development characteristics are similar to those of Western cities, they still have the features of Chinese cities themselves. The driving force of the initial urban emergence in China was agriculture. Around 8000 B.C., grains and rice were found in the middle reaches of the Yellow River and the Yangtze River, and they symbolise the local dynamics of local origins and agricultural development. The driving force of the initial urban emergence in China was agriculture. In the Middle Neolithic, around 6000–5000 B.C., a settled life based on farming emerged in China. Having a large number of settlements in the basins and plains, they were still spreading at that time. From 4000 to 2800 B.C., agriculture and handicrafts were further developed, contributing to the transformation of society at that time, and a primitive civilisation gradually emerged. Human organisation and social structure became complex, ruled by an elite class that held power. Until the late Neolithic period, the size and number of settlements increased more than tenfold because of technological and social changes, and primitive cities gradually formed. These cities were bastions of the elite ruling class of the time, who built walls and owned a moat. As of the small size of these cities, they could not accommodate entire tribes or entire groups, thus creating a difference between urban and rural areas with the surrounding villages, such as Chengtou Mountain in Hunan Province. At this time, the primitive cities had some order and planning. From 3000 to 2200 B.C., the social characteristics of the Longshan period emerged in different regions of China, including the development of handicrafts and religious practices, the development of agriculture and the formation of classes and the emergence of cities.

However, the region's development is inseparably linked to the city, and some scholars believe that the city represents a special region and is the centre of areas of different sizes. Urban Science, which treats urban areas as objects of study, is a branch of regional science. Cities are the basic constituent units of regions. The region and the city are a relationship between the whole and the part, influencing each other. A region is a contiguous geographical space where somehow there exists a unified economic and social system. From the perspective of regional science, it integrates theories and methods from multiple disciplines, such as economics, political science, architecture, transportation science and urban planning (Yang, 2019).

Regional spatial structure types are divided into rural and urban areas, with rural areas being large in scope and dominated by agricultural production activities. Non-agricultural activities dominate urban areas, and it also has the function of driving rural areas. From the characteristics of regional spatial layout, the regional transportation routes are shown as lines and networks, cities as points and urban clusters as islands. This chapter discusses the implementation of synergistic regional planning mainly in urban agglomeration mode. The origin of the region from the perspective of urban agglomeration can be traced back to the concept of the 'Garden City' proposed by Howard, a British social activist, in 1898. The 'Garden City' consists of two parts: the city and the countryside, with the city at the centre and a park at the centre of the city. Its six main roads radiate outward from the centre, dividing the city into six districts, and the outermost circle of the city is built with factories, warehouses and markets. Transportation is convenient with the outermost ring road on one side and a circular railroad spur on the other. When the number of people exceeded the capacity of the 'Garden City', Howard believed that over time the 'Garden City' would form an urban agglomeration, transforming into a complex of central social cities in the context of the environment where the subway had already appeared. Several garden cities surround the central city, which together forms a cluster of cities with agricultural dividers, and this central city will be larger. The cities have intertwined radial roads and intermunicipal railways. There are radial roads between the central city and each of the garden cities; on top of them, there are underground railways and circular intermunicipal canals. The extensive canal traffic line along the edge of the central city radiating to each garden city is accessible to the sea. The colonial cities are connected by transportation, water supply and drainage facilities to form a whole system. This urban agglomeration was described by Howard as a 'ghetto-free, smog-free urban agglomeration'. In 1899, Howard founded the 'Garden City' Association, and he was an active promoter of social reform. In 1903, the first generation of the 'Garden City' practice emerged in the London suburb of Letchworth. This was followed by a second generation of 'Garden City', which was also called the satellite town in Welling. Regional planning was formally introduced in 1915 by Patrick Geddes, who was one of the pioneers of regional planning theory. In his book 'Cities in Evolution: An Introduction to the Town Planning Movement and the Study of Civics', he proposed that the emergence of new technologies (electricity generation, the internal combustion engine) was leading to the evacuation of large cities and the formation of clusters. These clusters of regions and cities were called 'Conurbations'. He predicted the emergence of a megacity belt in Europe and the United States (U.S). Half a century later, his theory influenced Jean Gottmann's study of megalopolis zones (Hall, 1978). In 1933, German geographer W. Christaller proposed the theory of central location, which is the centre of the surrounding area; it is the centre of a city or a large gathering of residents, commerce and services. This theory defines the spatial organisation and structure of cities and urban agglomerations. It gradually developed into a basic theory for regional development and analysis (Fang & Yu, 2017). In 1957, French geographer Jean Gottmann published the study 'Megalopolis: or the Urbanisation of the Northeastern Seaboard'. The object of his study was the widely distributed contiguous megalopolis located along the northeastern coastline of the U.S. It is a cluster of large numbers of people, industrial and commercial facilities, financial wealth and cultural activities that no place can compare. Jean Gottmann argues that urban agglomerations are formed via the role of agglomeration and develop in the form of a network structure. These regions develop around a strong urban nuclear energy source (Gottmann, 1957). At the same time, the author mentions two factors that contribute to the contiguous development of urban agglomerations: firstly, the polynuclear origin, and secondly, the role of the 'hinge'. The authors provide an explanation for the identification of the role of the 'hinge'. He gives the example of the U.S. East Coast, which has assumed the role of a window for developing overseas relations, serving as a springboard for settlement and development in the interior. Whether the U.S. economy is moving overseas or shifting inland, the East Coast has always held the primary position. The U.S. also has the North American Great Lakes City Clusters and the San Diego–San Francisco City Clusters on the Pacific Coast of the south-western U.S. Other countries have formed several of the same city cluster regions, including the London City Cluster in the UK, the Paris City Cluster in France, the Ruhr City Cluster in Germany, the Randstad City Cluster in the Netherlands and the Pacific Coast City Cluster in Japan. Firstly, the emergence and development of foreign urban agglomerations are based on industrialisation and the division of labour amongst cities, which develop in the modes of core cities driving surrounding cities or multi-centre synergy. Both modes take advantage of each city's resources to maximise the region's overall benefits, for example, economic, political, high-tech and foreign trade resources. In addition, they focus on the linking function of a well-developed transportation network. Cooperation amongst cities in urban agglomerations and the formation of a complete system must rely on a well-developed transportation network. Most foreign city clusters have well-developed regional transportation infrastructure networks. Amongst them, well-developed railway and highway facilities constitute the skeleton and linking hub of the spatial structure of foreign city clusters. Secondly, they also focus on the coordination role of government, and secondary cities focus on dislocated development with core cities (Wang, 2005).

We want to explain here a concept of synergy, which was introduced in Hermann Haken's book 'University Translation Series–Synergetic: The Mystery of Nature's Composition', which was published in 1969. He formally created the concept of synergy, and the author hoped to find common principles that would apply to very different fields of science and that the theory of synergism would likely appear in different disciplines. Synergy means coordinated cooperation, and it is derived from the Greek language. Synergy was created to solve the problem of how parts form a whole via synergistic collaboration (Haken, 1969), and it explains the relationship between the whole and the parts. This principle plays the same supporting role for the theory of synergistic regional development. In the 1980s, Jean Gottmann's concept of the urban agglomeration concept to China in the book 'Introduction to Urban Geography' by Ning Yuemin, which has been influencing Chinese urban agglomeration planning. The driving force behind urban clusters and regional synergies in China is the need for such a synergy between the national economy and the country's overall competitiveness.

The 11th Five-Year Plan for National Economic and Social Development of the People's Republic of China proposed that urban agglomeration is the main form to promote China's new urbanisation. It will gradually form with the coastal and Beijing–Guangzhou and Beijing–Harbin lines as the vertical axis, the Yangtze River and Longhai lines as the horizontal axis, multiple city clusters as the main body, other cities and small towns distributed in a dotted pattern, and permanent arable land and ecological functional areas spaced apart. In this manner, an efficient, coordinated and sustainable spatial pattern of urbanisation can be created. China has already formed city clusters, such as Beijing–Tianjin–Hebei, Yangtze River Delta and Pearl River Delta regions, and they continue to play a driving and radiating role in strengthening the division of labour and cooperation amongst cities within city clusters and play a complementary and advantageous role to enhance the overall competitiveness of these regions. Regions with conditions for urban cluster development should strengthen integrated planning, with mega-cities and large cities as the leading cities and playing the role of central cities. Multiple new city clusters with less land, more employment, strong factor gathering capacity, and reasonable population distribution will be formed (State Council, 2006). China has approved nine city clusters, including the Yangtze River Midstream City Cluster, Ha–Chang City Cluster, Chengdu–Chongqing City Cluster, Yangtze River Delta City Cluster, Central Plains City Cluster, Beibu Gulf City Cluster, Guanzhong Plain City Cluster, Hubao–Egyu City Cluster and Lanxi City Cluster. During the 13th Five-Year Plan period, China proposed a strategic path of spatial integration of urban agglomerations to drive synergistic regional development, such as the Yangtze River Delta Integrated Development Strategy, which has achieved remarkable results. In the 14th Five-Year Plan period, China will continue to deepen and improve the overall regional development strategy based on the new type of urbanisation and endeavour to design new strategic paths for the synergistic development of urban agglomerations in the four major regions (Wei & Li, 2020).

The above review is a chronological review of the theories and concepts of urban origins and evolution, development from cities into regions, and regional synergy in China and the West. As Patrick Geddes stated, cities will develop through five stages, the second and third of which is the gradual aggregation of cities into a large urban region. Urban agglomerations are the result of the evolution of urban spatial forms. In the current environment of economic globalisation and regional integration development, often city clusters are the main arena of competition between countries and countries and between regions, and synergistic regional development is an inevitable trend. It reduces the gap between regions and improves regional competitiveness via the association of multiple cities in economic, resource and geographic space. It is a process in which two or more cities or urban functions achieve their goals by means of spatial aggregation. These cities share common interests and common destinies. At the same time, the strategic decisions of countries in different periods will influence the synergistic regional development into a new stage, and synergistic regional development has become an important research theme in various fields. In the next section, we review and comment on domestic and international cases of synergistic regional development.

2.2 Cases and Roles

The purpose of synergistic regional development is to achieve sustainable regional development and enhance regional competitiveness. There have been many domestic and foreign cases of synergistic regional development, which provide a rich experience for our strategic practice of synergistic regional development. Then, we elaborate on the synergistic development of the cases in terms of the two aspects of industrial layout and transportation. Finally, we analyse and summarise the domestic and international cases.

The industrial layout is the adjustment, overall layout and planning of the industrial structure. The structure of the region includes the regional industrial structure and the regional spatial structure. The regional industrial structure is also an essential part of promoting regional development, which is influenced by natural geographical conditions, economic development level, resource allocation status and labour quality. The economy is one of the factors affecting the industrial structure, and urban development planning cannot be separated from the urban economy. At this stage, major Chinese cities propose to optimise the industrial layout, which is one of the paths to promote regional industrial synergy. Meanwhile, strengthening transportation infrastructure is also an important element of synergistic regional development.

Take the Yangtze River Delta City Cluster, which is more mature in synergistic development in China, as an example. The development model of the Yangtze River Delta City Cluster is based on the national central city as the core. From an economic perspective, the Yangtze River Delta urban agglomeration is one of the most dynamic, open and innovative regions in China. It is also an important intersection of China's Belt and Road Economic Zone and the Yangtze River Economic Zone. The Yangtze River Delta City Cluster is further developed by means of a higher level of international cooperation and competition. It plays an essential role in supporting and leading China's economic and social development. The Yangtze River Delta City Cluster mainly includes cities in Shanghai, Jiangsu, Zhejiang and Anhui provinces and other cities within these four provinces. With Shanghai as the core, several cities are linked to form a cluster. China's planning period for this region is 2016–2020, with a long-term outlook to 2030. In the direction of synergistic development of industrial layout and transportation facilities construction, the document 'Yangtze River Delta City Cluster Development Plan' clearly proposes to promote the synergistic development of Shanghai and neighbouring cities, such as Suzhou, Wuxi, Nantong, Ningbo, Jiaxing and Zhoushan, to play a guiding role for the integrated development of the Yangtze River Delta City Cluster and to enhance the ability to serve national strategies, such as the Yangtze River Economic Belt and the 'Belt and

One Road', to build a transportation infrastructure with reasonable layout, perfect function, safety and efficiency whilst improving the interconnection of transportation facilities, establish a comprehensive transportation network mainly based on railway transportation and improve the intercity comprehensive transportation network. On the basis of the national comprehensive transportation corridor, with Shanghai as the core and Nanjing, Hangzhou and Hefei as the subcore, build a multi-level integrated transportation network mainly with high-speed railroads, intercity railroads, expressways and the Yangtze River Golden Waterway (National Development and Reform Commission, 2006). The Yangtze River Delta City Cluster has also established the Yangtze River Delta Regional Cooperation Office, which is able to participate in the overall management and supervision of implementation actively and can better actively guide enterprises and various sectors of society to join the cooperation.

The total planned area of Shanghai Hongqiao Business District is 86.6 km^2 to build an international open hub leading the integrated development of higher quality in the Yangtze Delta region. Through this business district, a driving effect on the surrounding areas will be formed, eventually forming a unique integrated business district.

The construction planning of Suzhou High-Speed Railway New City in Jiangsu Province started from the development of Suzhou Xiangcheng District and the strategy of expanding the northern part of Suzhou city, which is positioned to build a national high-speed railway hub and a hub of the intercity railroad network in the core region of Yangtze River Delta. It forms a 'double cross' transportation hub with the Nantong–Suzhou–Jiaxing–Ningbo high-speed railway line, Suzhou and Huzhou intercity railroad, and Suzhou, Wuxi and Changzhou intercity railroad intersecting at Suzhou North Station. At the same time, it deepens the strategic cooperation with Shanghai Hongqiao Hub to create an integrated hub of Shanghai and Jiangsu. The industrial planning of this area is positioned as the new gateway to Suzhou, the new home of the city, the new industrial highland and the new ecological space. With global services as the scope, it will innovate the process of one-stop trade services and create a trade service platform based on the Yangtze Delta region and connected to the world via cross-border cooperation. It uses the Yangtze River Delta International Research and Development Community to develop industry–academia–research activities with Shanghai universities, including big data, industrial Internet, financial technology and other aspects, to achieve the transformation of 'science' to 'technology' to 'industry' and improve the regional innovation chain in this manner. It has built a film and television industrial park, with film and television, exhibition, e-sports, games and creative design as the main industry, with the aim of cultivating professionals in these fields by means of the form and function of the industrial park and improving the living facilities of the region. Moreover, this area relies on the medical resources of the Chinese Academy of Medical Sciences, the University of Washington School of Medicine and Johns Hopkins University School of Medicine to jointly build an international medical and health centre, which strengthens the synergistic innovation with Shanghai medical technology and cultivates medical talents through a transnational joint manner. Summarising the above point, the trade service platform industry, high-tech industry, cultural and creative industry and medical and healthcare industry in Suzhou, Jiangsu Province's high-speed railway new city, have strengthened and promoted the regional industrial synergy development via cross-border cooperation, integration of industry, academia and research for industrial aggregation.

Driven by market-oriented mechanisms, the new high-speed railway city in Jiashan County, Jiaxing City, Zhejiang Province, was developed in integrated public–private partnership mode. It identifies development positioning, such as a new centre for urban development, which is a significant platform for industrial innovation, and the creation of industrial clusters, such as life medical care, intelligent automobiles, business services and film and media. At the same time, it strengthens synergistic cooperation with Shanghai. It takes over the transfer of talent resources from Shanghai to Jiashan County based on the talent entrepreneurship park in Shanghai. Shanghai is the beginning of the development of this region, and the process of this region's development is in Jiashan.

With the official opening of Bengbu South Station of the Beijing–Shanghai High-Speed Railway in 2011, the planning and construction of Bengbu High-Speed Railway New City in Anhui Province began, with a total area of 41 km^2. Its regional planning is positioned as a bearing area for the implementation of Bengbu's 'two centres' strategy, a pilot area for the integration of Bengbu City, Huaiyuan County, and Fengyang County, a leading area for modern city construction, a cluster area for medium and high-end modern service industries and business support, and a new highland for science education, innovation and entrepreneurship. In 2020, Bengbu City High-Speed Railway New City was again upgraded industrially, mainly positioned for the development of science and innovation industries. According to the actual development of Bengbu High-Speed Railway New City, the way of industrial structure adjustment in the early stage is the mode of industrial gathering, and the mode of industrial upgrading in the later stage, which strives to form a regional industrial gathering place with certain influence via the method of regional industrial structure adjustment. At the same time, Bengbu High-Speed Railway New City takes Bengbu West Station as the core for spatial planning, integrating public transportation resources and forming a regional comprehensive transportation hub.

The Northeast Atlantic Coastal City Cluster of the U.S. is a contiguous metropolitan area that includes Boston, New York, Philadelphia, Baltimore and Washington. It spans 12 states and 1 special district. The Atlantic Coastal Cities cluster in northeastern U.S. has developed a core system of knowledge-intensive industries, such as finance and information. City government agencies and civic groups largely coordinate this urban agglomeration. The northeastern Atlantic coast of the U.S. is mainly dominated by the financial and business services industry and is called the 'Financial Bay Area', based on the Port of New York, which has gradually become the core of the U.S. economy. In 2016, the Northeast Atlantic Coast metropolitan area of the U.S. was the top two in terms of GDP for real estate and insurance, and it has a tertiary sector share of more than 90%. Furthermore, major world financial and securities firms are clustered here, creating an excellent financial industry base. The structure of the Northeast Atlantic Coast of the U.S. is hierarchical, with the first tier of cities being New York, the second tier being Boston, Philadelphia, Baltimore and Washington, and the third tier comprising approximately 40 other small- and medium-sized cities. New York, Boston, Philadelphia, Baltimore and Washington have their own strengths and different leading industrial resources. New York is represented by financial and business industries, high-tech industries are concentrated in Boston, and industries, such as clothing, cosmetics, printing and oil, military and metal products manufacturing, are focused in the area around Manhattan. The defence, aviation and electronics industries are gathered in Philadelphia, and the mining and shipping industries are gathered in Baltimore. These five cities implement a strategy of dislocated development and complementary resources, and New York and its surrounding cities have formed a pattern of synergistic industrial development with diversified, comprehensive and complete industrial chains, which provides the foundation and guarantee for the industrial development of the Atlantic Coastal City Cluster in northeastern U.S. On the basis of the spatial and industrial structure, the transportation system of northeastern U.S. urban agglomeration is gradually integrated. The air transportation industry in the Northeast U.S. City Cluster is highly developed, and the three airports of JFK, Newark and LaGuardia constitute a world-class aviation corridor. The airport construction is organically integrated with the ground highways and railroads, maximising the efficiency of air transportation. Transportation within the urban agglomeration is dominated by highways and railways, with railways bearing the majority of passenger traffic. Passenger and cargo transportation within the city limits is dominated by road car transportation. New York City has a special system of passenger rides with in-station transfers at the entrances to the city. This measure enables integrated transfer services between private cars and buses, road transportation and rail transportation. It enhances the convenience of transportation within urban clusters and promotes the two-way flow of resource factors.

The Rhine–Ruhr urban cluster in Germany has one city of one million people and nearly 30 small- and medium-sized cities. The Rhine–Ruhr region is rich in coal and water resources. It has convenient water and land transportation conditions. In the 1950s, to reduce the logistics costs of importing mineral resources, the German Rhine–Ruhr City Cluster underwent an industrial transfer, with steel companies moving from the inner cities to the port cities. Duis is a port city, and this area is mainly clustered with heavy industries as leading industries, such as logistics, steel and petrochemicals. The headquarters economy of the Rhine–Ruhr City Cluster is in Cologne, where the leading industries are insurance, media, conventions and headquarters economy. In Düsseldorf, there are leading industries, such as communication, advertising, finance, convention and exhibition and the headquarters economy. The cities of Cologne and Düsseldorf have the advantage of transportation hubs. In the 1960s to the 1980s, high-end elements were concentrated in Cologne and Düsseldorf to share with surrounding cities, creating cloud and chain synergy. The high-tech industry was concentrated in Dortmund, which included new and high-tech industries, such as electronic information and biological hospitals. In the 1980s, core resources in the city cluster were integrated. Dortmund establishes technology centres with Essen and Duisburg. Companies and research institutes cooperate to create a technological path in the Ruhr. In the Rhine–Ruhr City Cluster, information and talent assist the tertiary sector and technology in collaborating within the city cluster by means of cloud-like synergies, promoting the development of new industries and the formation of industrial chains.

The Ruhr region has a well-developed transportation network, and it is the intersection of Europe's major east–west and north–south transportation routes. It has the largest river port in the world, busy inland waterways and the densest rail network. Dusseldorf has the third largest international airport in Germany, and the region is extremely convenient for highways. The different modes of transport use their strengths to form a unified and coordinated integrated transport system in the Ruhr. Thus, the Rhine–Ruhr urban agglomeration has a well-developed regional public transport system with rail as the backbone, based on good transport infrastructure. The railway is operated in verkehrsverbund mode. The railway is planned, managed and operated by the Verkehrsverbund Rhine–Ruhr, which is the coordinating organisation between the government and the specific operating companies. The railway network skeleton includes high-speed railroads, suburban railroads, subways, light railways and trams. It connects large and small cities in urban agglomerations. It is characterised by punctuality, speed, comfort and safety. The Rhine–Ruhr urban agglomeration is highly integrated via the interconnection of different modes of the railway. Installing safe and quick mini-transfer feed stations is an easy transfer between rail transit and buses, bicycles and private cars. Bus stops, secure bicycle storage and private car-parking have been set up outside subway stations. These measures have effectively solved the problem of the 'last mile'. Finally, Germany has adopted legislation to ensure intercity cooperation in transport. This scenario provides for efficient and convenient transport services to the greatest extent possible.

There are three metropolitan areas in the Pacific City Cluster of Japan, and they are Tokyo–Yokohama, Osaka–Kobe and Nagoya urban areas. These metropolitan areas emerged because of the opening of Japan's Shinkansen, which made it more convenient for the population and capital to flow between metropolitan areas. In 1956, Japan proposed the construction of a 'metropolitan area' with Tokyo as the centre and a radius of 100 kms in the document 'Metropolitan Area Preparation Law'. The Pacific City Cluster of Japan also covers Kanagawa, Saitama, Chiba, Ibaraki and other prefectures. The Japan Pacific Urban Agglomeration adopts a core city-oriented mode. At the same time, it has been continuing the planning goal of multiple cores dispersed. The 23 special districts of Tokyo are the core of the mid-level urban agglomeration, which drives the development of Tokyo's peripheral cities. Ibaraki and several prefectures form the outermost layer of the urban agglomeration. The cities cooperate and develop together. Japan's Pacific City Cluster has established a top–down authority to guarantee the synergistic development between cities. Its experience of coordination comes from the development and implementation of their planning activities. The government coordinates between cities by using unique plans for industrial policy, regional functional division of labour and the natural environment.

The Pacific City Cluster of Japan has a large industrial cluster. Within the urban cluster, the Keihin and Keiyo areas are two major world-class industrial zones, concentrating major industrial sectors of traditional heavy industry and modern high-end manufacturing. It is home to one-third of Japan's population and contains one-third of its total economic output and 40% of its industrial output. It also has a solid industrial base of about 60 Fortune 500 companies, and the Japan Pacific City Cluster is home to financial, research and development, cultural, and entertainment industries. Therefore, the formation of economic unification in the Keihin and Keiyoha regions will enhance competitiveness. The critical factor in their unification is the unified planning layout and the clear division of functions of each port. They have classified the status and level of ports according to their geographical location and throughput capacity, exploited the resources and advantages of each port in the bay area, encouraged healthy competition within ports and promoted the synergistic development of regional industries. Regarding transportation synergy, the transportation integration of the Pacific City Cluster in Japan is highly rated. The reason is that it has a fast, reliable and safe urban transportation system. The Japan Pacific City Cluster has the largest port cluster and aviation network, and the external transportation for the entire city cluster is mainly by water transport, air transportation and railway. The traffic within the urban agglomeration is primarily by road and railroad, and the passenger capacity of the railroad is 85% of the total number of passengers, which proves that the railroad transportation within the urban agglomeration is well developed. On the basis of the multi-level spatial structure of the Pacific City Cluster in Japan, the intercity railroads and other rail transportation in the city cluster have developed into a multi-level and multi-class integrated transportation network, which plays the role of intercity public transportation in the city cluster. At the same time, the highway construction layout is 'three rings and nine radials', which forms a three-dimensional interactive, integrated transportation network with railway transportation.

In summary, the world-class city clusters developed by different countries according to their development environment with different synergistic development modes. In terms of the city cluster synergistic development mode, China's Yangtze River Delta City Cluster is a mode with the national central city as the core. There are two types of city cluster synergistic development modes in foreign countries, the central city-based mode and the multi-centre synergistic mode. For example, the Northeast Atlantic Coastal City Cluster of the U.S. and the Pacific City Cluster of Japan are based on the central city mode. At the same time, the Rhine–Ruhr City Cluster is a multi-centre synergistic mode. In summary, from both industrial layout and transportation, there are still differences between the regions of China's Yangtze River Delta City Cluster. Currently, the industrial structure optimisation and transportation integration construction of the Yangtze River Delta City Cluster is in the process of further enhancement.

By contrast, urban agglomerations in the U.S., Japan and Germany tend already to have mature spatial structures and mature industrial structures. They concentrate on the advantageous industries of each city within the urban agglomerations, actively participate in global competition and have high economic connectivity within the urban agglomerations. Their transportation infrastructure networks have developed and have a perfect transportation integration system. Their excellent transportation infrastructure networks provide the basis for economic and social development. Thus, the urban agglomerations in developed countries are at a high level of urbanisation.

The synergistic development experiences of the developing Yangtze River Delta City Cluster and international city cluster have certain implications for other city clusters. In the next section, we take the Hengqin–Guangdong–Macao Deep Cooperation Zone as an example and analyse it. The advantages and difficulties in forming the synergistic development of the Hengqin–Guangdong–Macao Deep Cooperation Zone are discussed in comparison with the regional synergistic development experiences of domestic and foreign cases.

2.3 Hengqin and Macao Synergy: Advantages and Difficulties

In the context of the development and construction of the Guangdong–Hong Kong–Macao GBA, the document 'Outline of the Development Plan of Guangdong–Hong Kong–Macao Greater Bay Area' points out that relying on the advantages of Hong Kong and Macao as free and open economies and Guangdong as the leader of reform, it builds a platform for cooperation and development amongst Guangdong, Hong Kong and Macao. The purpose of building the Hengqin Guangdong–Macao In-Depth Cooperation Zone is to promote the development of the Guangdong–Hong Kong–Macao GBA. The Hengqin Guangdong–Macao In-Depth Cooperation Zone is an elaborate new model, new platform, new demonstration and new highland. This is a major deployment to enrich the practice of 'One Country, Two Systems' (Central Committee of the CPC and the State Council, 2019). In September 2021, the document 'Overall Plan for the Construction of Hengqin Guangdong–Macao Deep Cooperation Zone' issued by the Central Committee of the Communist Party of China and the State Council pointed out that Hengqin is adjacent to Macao and has the inherent advantages of Guangdong–Macao cooperation. It should deepen the cooperation amongst Guangdong, Hong Kong and Macao and promote mutual benefits, common development and progress amongst cities in the Guangdong–Hong Kong–Macao GBA (Central Committee of the CPC and the State Council, 2021). At the same time, this is also the process of coexistence of opportunities and challenges.

From the perspective of China's overall development pattern, China will build a new development pattern during the 14th Five-Year Plan period. From the perspective of China's overall development pattern, a new development pattern will be built during the 14th Five-Year Plan period. The document 'Proposal of the Central Committee of the Communist Party of China on the formulation of the 14th Five-Year Plan for National Economic and Social Development and Vision 2035' mentions maintaining Macao's long-term prosperity and stability and supporting the SAR in consolidating and enhancing its competitiveness. To achieve diversified and sustainable economic development and to help Macao better integrate into the national development pattern (Central Committee of the CPC, 2020). The integration of Macao into the national development pattern is a significant opportunity to achieve sustainable development. In terms of Macao and Hengqin's strengths, Macao is a crossroads for domestic and international exchanges. It also has the advantages of tourism, technology research and development, multi-culturalism and a Sino–Portuguese intermediary platform. Macao's overall tourism industry is performing well, driven by the gaming tourism industry. Every year, Macao hosts international tourism events, such as the Grand Prix, Macao International Music Festival and other events that attract many international friends and visitors to Macao. With the strong support of the State and the Macao government, Macao has also been able to achieve good results in scientific research and development. Macao already has four State Key Laboratories, and Macao is the only city on the west bank of the Pearl River with State Key Laboratories. In the sixteenth century, Macao was an important stronghold of the Maritime Silk Road and the earliest base for two-way cultural exchange between the East and the West. Moreover, under the influence of historical, cultural, linguistic and human relations factors, Macao has always had traditional and extensive ties with Portuguese-speaking countries. Given this environment, Macao has also been optimising the foundation and conditions of its service platform for economic and trade cooperation with Portuguese-speaking countries. Under the impact of the novel coronavirus pneumonia, the limitations of Macao's development have been exposed more clearly. Regarding problems of having a small space or single industrial structure, it has hindered Macao society from maintaining long-term sustainable growth. Hengqin will play a core engine function in the construction of the entire Guangdong–Hong Kong–Macao GBA, leading and driving. Its role is mainly to support and serve Macao. Hengqin is adjacent to Macao and has spatial advantages as a newly developed area with a land area of approximately 106 km^2. It can effectively alleviate the current situation of high population density in Macao. Combining the two places' advantages and promoting their synergistic development can provide new space for Macao, enrich Macao's industries and enhance the regional competitiveness of the west bank of the Pearl River Estuary.

As the construction of the Hengqin–Guangdong–Macao Deep Cooperation Zone is under 'One Country, Two Systems', two customs zones and two currencies, the difficulties of building the Hengqin–Guangdong–Macao Deep Cooperation Zone under the mechanism of codevelopment, comanagement and sharing cannot be ignored. It must deal with the obstacles brought by the difference between 'two systems'. There are still gaps in the interface between the systems of Macao and Hengqin. The cooperation between Macao and Hengqin has entered into deeper institutional issues, which require a deeper level of breakthrough in the barriers to relevant policies. Compared with world-class city clusters, the sharing mechanism between the two places is not established. For example, how will the results of cooperation be coordinated during the synergistic development of Macao and Hengqin? How will the outcome of Hengqin complement the moderate diversification of Macao's economy? These questions are still in the stage of exploration and research. The development of industrial synergy between Hengqin and Macao has not reached the expectation, and the industrial connection between the two places has been hindered. In industrial planning, Hengqin is set to be a high-technology and modern service industry with high technical requirements and capital needs, resulting in a high threshold for pre-investment projects. It is difficult for small- and medium-sized enterprises in Macao to join the cooperation zone.

Meanwhile, Macao's tourism industry is an advantage. Nonetheless, it operates separately from Hengqin's tourism industry, and the differences in the tourism industry between the two places are more significant than the fusion. Referring to the experience of industrial synergy development of the city cluster in the case, for Hengqin and Macao to achieve practical complementary mutual assistance and synergy development of industries, the overall mechanism of synergy and cooperation between the two places should be clarified. The two places should jointly study industrial planning and apply industrial policies for synergy between the two places. According to the actual situation of industrial development of the two places, a fine selection of industries to be cultivated. To give full play to the industrial advantages of the two places and to clarify the division of labour and collaboration between the two cities. Let Hengqin and Macao industries experience the effect of complementary resources and adhere to the principle of industrial differentiation, the implementation of industrial dislocation development. Avoid the consequences of industrial homogenisation between the two cities and low-level competition. In addition, the high-quality synergistic development of world-class city clusters is due to the focus on the communication role of a well-developed transportation network. The development of transportation and information industries is an essential condition that contributes to the rapid development of city clusters. The Macao Light Rail Extension Hengqin Transportation Line project is officially underway. It is expected that by 2025 at the earliest, people will be able to reach Macao directly by light rail in Hengqin. Regarding high-speed rail and intercity transportation, Macao will also cooperate in promoting the planning and construction of projects, such as the Guangzhou–Zhuhai (Macao) High-Speed Rail and Nansha–Zhuhai (Zhongshan) Intercity Railroad, which are the critical work plans in Macao's transport planning long-term tasks for 2030.

The integration of transportation between Hengqin and Macao is in its infancy, and solving the 'last mile' of urban transportation in the future is a difficult subject. Nonetheless, it is essential to solving this problem. It can effectively improve the efficiency of transportation services. To achieve synergy and cooperation between Hengqin and Macao, a well-developed transportation network must be built, and Hengqin and Macao need to accelerate the completion of transportation integration.

In a word, the synergistic regional development in this chapter mainly refers to the regional cooperation formed in the form of city clusters. In an era of increasing global connectivity, synergistic regional development is already an essential urban management and planning theme and will see significant progress in the future. On the one hand, synergistic development mechanisms between regions are becoming more complete. On the other hand, cities worldwide strive to become more sustainable and resilient. Thus, the theory and practice of regional synergy provide an essential opportunity to help understand how systems between regions work together. It can provide better design and governance policies for cities and regions.

References

Fan, H. S. (2011). Theory and practice of promoting coordinated regional development in China. *Comparative Economic and Social Systems, 06*, 1–9.

Fang, C., & Yu, D. (2017). Urban agglomeration: An evolving concept of an emerging phenomenon. *Landscape and Urban Planning, 162*, 126–136. https://doi.org/10.1016/j.landurbplan.2017.02.014

Gao, C. H. (2016). Urban-regional justice and synergistic development of urban agglomerations—A perspective of contemporary Western urban theory. *Journal of Jiangxi Normal University (Philosophy and Social Science Edition), 04*, 36–40.

Gottmann, J. (1957). Megalopolis or the urbanization of the northeastern seaboard. *Economic Geography, 33*(3), 189–200.

Haken, H. (1969). University translation series—synergetic: The mystery of nature's composition. English edition: Haken, H. (2005). University translation series—synergetic: The mystery of nature's composition (trans: Ling Fuhua). Translation, Shanghai.

Hall, P. (1961). *Cities of tomorrow: An intellectual history of urban planning and design since 1880.* MIT, New York. English edition: Hall, P. (1978). *Cities of tomorrow: An intellectual history of urban planning and design since 1880.* (trans: Tong Ming) Tongji University, Shanghai.

Mumford, L. (1961). *The city in history: A powerfully incisive and influential look at the development of the urban form through the ages.* English edition: Mumford, L. (2005). *A powerfully incisive and influential look at the development of the urban form through the ages* (trans: Song Junling, Ni Wenyan). China Construction Industry, Beijing.

National Development and Reform Commission. (2006). *National development and reform commission ministry of housing and urban-rural development notice on the issuance of the Yangtze river delta city cluster development plan.* Retrieved June 30, 2022, from https://www.ndrc.gov.cn/xxgk/zcfb/ghwb/201606/t20160603_962187.html?code=&state=123. Accessed 1 June 2022.

The Central Committee of the CPC and the State Council. (2019). *The Central committee of the communist party of China state council issued the outline of the development plan Of Guangdong, Hong Kong and Macao greater bay area.* Retrieved June 30, 2022, from http://www.gov.cn/zhengce/2019-02/18/content_5366593.htm#1. Accessed 10 June 2022.

The Central Committee of the CPC and the State Council. (2021). *China issues plan for building Guangdong-Macao in-depth cooperation zone.* Retrieved June 30, 2022, from https://english.www.gov.cn/policies/latestreleases/202109/05/content_WS6134a98ec6d0df57f98dfb96.html. Accessed 10 June 2022.

The Central Committee of the CPC. (2020). *Proposal of the central committee of the communist party of China on formulating the 14th five-year plan for national economic and social development and the visionary goals for 2035.* Retrieved June 30, 2022, from http://www.xinhuanet.com/politics/2020-11/03/c_1126693293.htm. Accessed 10 June 2022.

The State Council. (2006). *The state council on the implementation of the "National economic and social development of the People's Republic of China and social development of the 11th five-year plan notice on the division of work on the main objectives and tasks.* Retrieved June 30, 2022, from http://www.gov.cn/gongbao/content/2006/content_413969.htm. Accessed 1 June 2022.

References

Wang, N. J. (2005). A new exploration of the development model and experience of foreign urban agglomerations. *Research in Technology Economics and Management, 02*, 83–84.

Wei, H. K., Nian, M., & Li, L. (2020). China's strategies and policies for regional development during the period of the 14th five-year plan. *Chinese Journal of Urban and Environmental Studies, 8*(2), 2050008. https://doi.org/10.1142/s2345748120500086

Yang, K. Z. (2019). New urbanization and coordinated regional development. *Chinese Journal of Urban and Environmental Studies, 7*(4), 1975009.

Open Access This chapter is licensed under the terms of the Creative Commons Attribution 4.0 International License (http://creativecommons.org/licenses/by/4.0/), which permits use, sharing, adaptation, distribution and reproduction in any medium or format, as long as you give appropriate credit to the original author(s) and the source, provide a link to the Creative Commons license and indicate if changes were made.

The images or other third party material in this chapter are included in the chapter's Creative Commons license, unless indicated otherwise in a credit line to the material. If material is not included in the chapter's Creative Commons license and your intended use is not permitted by statutory regulation or exceeds the permitted use, you will need to obtain permission directly from the copyright holder.

Chapter 3
Cooperation Planning System for Hengqin and Macao

Abstract The planning system, as a crucial instrument for the government to lead spatial development, is differently designed and operated in Hengqin and Macao within the 'One Country, Two Systems'. To ensure the cooperation of these two systems, experience from international and domestic cases as cross-border cooperation of planning will be analysed. Four aspects of the planning system, namely legal system, formulating plans, planning management and planning practice, are comparatively studied in Hengqin and Macao. The master plan is employed as an example to display the similarities and differences between planning systems in these two regions. From the analyses, this chapter proposes suggestions to contribute to future planning-led development in Guangdong–Macao In-Depth Cooperation Zone in Hengqin.

Keywords Planning system · Guangdong–Macao in-depth cooperation zone · Hengqin · Macao

3.1 Introduction on Planning Systems

Planning is a regulation for the government to guide and coordinate spatial development in terms of policy, strategies and projects; at regional level such targets are achieved by establishing spatial structure and principles in development and setting locations of regional infrastructure (Healey et al., 1997). Since the 1990s, the European Union (EU) has employed planning to cooperate between different spatial levels from social, economic and environmental dimensions. This type of cooperation aims to reduce spatial conflicts and uneven development within the EU and inspire other regions to develop a better interaction between nations, cities and communities (European Commission, 1999; Nadin, 2007). Hengqin is the site for Guangdong–Macao In-Depth Cooperation Zone, its planning system has the potential to guide and coordinate its spatial development, particularly to ensure cooperation between two planning systems, namely Mainland China and Macao. Within the background of 'One Country, Two Systems', the cooperation between Guangdong and Macao in Hengqin may be distinct from other parts of the world. The left part of this chapter will be organised as follows: the second part is case studies to provide experience for cross-border planning coordination; the third part concerns history and background for Hengqin–Macao coordination; the fourth part will introduce two planning systems and their potential conflicts; the fifth part employs the master plan as an example to illustrate the differences between two systems; the last part is suggestions for future coordination in planning.

3.2 Case Studies

3.2.1 Cross-Border Planning Coordination Within the EU

Given the significant disparity in development levels amongst EU member states and the gathering of multiple cultures in the region, the European Spatial Development Perspective (ESDP) is issued to avoid an increase in spatial development imbalances. Instead, the EU wants to form a common goal to guide the spatial development of each member state. Therefore, the ESDP was adopted in May 1999 due to the need for clear and forward-looking development guidelines at different spatial levels, as countries complement each other in their development.

The basic goal of ESDP is to strengthen the EU's socioeconomic cohesion and achieve balanced, sustainable development by means of coordinating economic, social, natural and cultural purposes. However, ESDP does not have a legal effect and is a guideline within the EU, which adopts a multi-level governance mechanism to break through the overall limits of the region. The core of its implementation is cooperation (European Spatial Planning Observation Network, 2007).

ESDP is led by the EU and implemented by means of sectoral and thematic areas, such as structural funds and grants, European transport network planning, aid programmes for cross-border regional cooperation and environmental policy. Cooperation projects are carried out voluntarily at the EU, transnational and local levels. The focus is on transnational cooperation, mainly via the Regional Cooperation Programme named INTRTTREG, which promotes project-oriented transnational cooperation (Zhang, 2011). ESDP, therefore, proposes a polycentric and balanced spatial development strategy, which is taking advantage of the coastal location, establishing a global economic integration region and an integrated transportation network, increasing accessibility, improving national/local connections and developing a polycentric urbanisation system and city clusters. Besides, ESDP proposes to promote the transformation of the rural economy from single to diversified, stimulate the potential of renewable energy, establish urban–rural partnerships, strengthen the cooperation between urban and rural areas and fully exploit the potential of local resources to reflect the uniqueness of the region and not to copy the development strategies of other regions.

3.2.2 Cross-Border Planning Coordination Between Germany and Poland

In this case, Germany is a long-established EU member state with a mature national planning system; and Poland emerges as a new EU accession country. Several cities are located on the border between the two countries: Oder–Neisse line, which has close historical and geographical ties, developed as 'laboratories' for German–Polish cross-border cooperation. In terms of the planning system, Germany is divided into two areas: the 'spatial order' system, which includes the national and state plans, and the 'building development directive planning', which consists of the plans of the local communities. When the two conflicts arise, the spatial order system has higher priorities. In Poland, the planning system is divided into three levels, namely the national, regional (voivodeship) and local (communal) levels, whilst the 'national spatial management concept' has a very vague function at the national and regional levels of territorial planning. Consistency is determined by mutual agreement in the cooperation between local planning and higher-level planning, and local governments are allowed to practise flexibly. In practice, Polish jurisdiction does not allow any plan provisions other than the commune's 'local spatial management plan' to be taken into account when granting or denying planning permission. Therefore, the local commune has a prominent position in the planning system.

In the 'European Garden 2003' cross-border planning concept proposed by Frankfurt-upon-Oder and Slubice, both cities wanted to create a public space on the banks of the Oder River to connect their urban green space networks. Nonetheless, such a concept did not define specific parameters or characteristics of the public space and was finally approved by both city councils as a conceptual strategy document. In practice, differences in the planning system and understanding of planning created a conflict in cooperation: as the project involved public land, the logic of the Polish planning system did not require preparing a 'local space management plan'. As for the informal concept document approved by the city council, the Polish planners would consider it as a guiding document without compulsory forces. So, the Polish side treated the joint plan area as a street-space upgrading project and distributed it to multiple contractors without concerted design and joint planning. In the German planning system, however, such documents are bound to the planner's practice. Thus, they have conducted joint planning for the whole project and updated their city master plan accordingly. The lack of joint planning of the public space on the Polish side led to strong dissatisfaction and concerns about the future on the German side. In this cross-border collaboration, the conflicts mainly stemmed from differences in planning systems and understanding. In a subsequent round of cooperation, the two sides defined the objectives and specific contents of the plan more precisely. They strengthened the communication between officials, local groups and individuals to reduce conflicts in cross-border planning (Tölle, 2013).

3.2.3 U.S.–Mexico Cross-Border Planning Coordination

A typical example of a transfrontier urban space is the border between the U.S. and Mexico. This area covers 2000 miles of the U.S.–Mexico border from Matamoros–Brownsville to Tijuana–San Diego, and more than 12 million people live in these cross-border cities. Both cities' social systems and ecosystems are integrated, and the First Worlds and Third Worlds intertwine here. The U.S.–Mexico border is addressed via the Border Liaison Mechanism, State Alliances and the Councils of Government, which brings national, state and city officials from both sides to address border issues here.

In the cross-border collaboration between Tijuana and San Diego, the two sides share a standard hydrological system (Tijuana River). However, their management systems and regulatory requirements are not the same, which led to sewage spills from Tijuana to San Diego, and this spilling has plagued the region for decades. This conflict was eventually resolved by the two sides working together to create an independent management authority. A successful example of cross-border cooperation in the region comes from Laredo–Nuevo Laredo, where a joint municipal plan for the Joint Urban Plan (La Carta Urbana de Los Dos Laredos) was developed by the U.S. and Mexican federal governments. Subsequently, the two sides conducted a joint environmental management plan and a joint historic preservation plan. Furthermore, joint actions in environmental protection, tourism development, transportation management and historical and cultural preservation successfully create a favourable political climate for cross-border plans and environmental management. Another conflict stems from the differences in legal systems and institutions between the two sides. Planning officials of the U.S. border cities claim that the six-year change in sessions of the Mexican government affects the planning of the border cities, hindering the two sides to continue the last round of collaborative events. At the local level, Mexican officials mostly come from political appointments and move on after accomplishing their own goals. These frequent changes in the collaboration between the two sides similarly affect the durability of collaborative planning. Two sides resolved such conflicts by establishing an independent administrative sector and developing joint urban plans. In addition, officials on both sides noted that informal, face-to-face and one-on-one interactions between the parties were the most effective approach of coordination. Many innovative local and informal arrangements in collaborative practice have been successfully applied in familiarising the parties with project issues and in implementing various tasks (Herzog, 2020).

3.2.4 Transboundary Planning Coordination Between Singapore and Johor, Malaysia

Since Singapore was independent of Malaysia in 1965, it has experienced rapid growth. However, its geographical location, in the south of a much larger country than itself—Malaysia, leads to a sensitive relationship with Malaysia (Zhou, 2004). Given the lack of space and resources, Singapore has shifted its investments to Johor, Malaysia, which is close to the territory and possesses abundant water resources and relatively cheap land. For instance, the Johor River is known as one of Singapore's 'Four National Taps' and accounts for 60% of Singapore's total water consumption (Chuah et al., 2018), South Johor is a fertile ground for Singapore to develop its business (semiconductor and other manufacturing activities) (Rizzo & Glasson, 2011). These factors have led to a strong transboundary relationship between Singapore and Johor. approximately 30,000–50,000 individuals cross the Straits of Johor to Singapore on weekdays to enjoy higher wages; conversely, on weekends, Singaporeans prefer to enjoy low-cost goods and entertainment in Malaysia in Singapore dollars. These cross-border dynamics have influenced many levels of local society, making Johor–Singapore a rapidly emerging transnational metropolitan area in Southeast Asia (Rizzo & Glasson, 2011).

The awkward challenge facing Johor in its transboundary collaboration with Singapore is lacking efficient public transportation, which can improve the daily commuting efficiency from Johor to Singapore. Johor's private bus-dominated transport is based on the idea of the 'most profitable route only' (i.e. from the centre of Johor to the causeway connecting Singapore to Johor). As a result, Johor residents employ a car or motorcycle-based transport system due to the lack of buses. In addition, the dual checkpoints in Malaysia and Singapore and the poor organisation on both sides of the Straits of Johor result in a 1.5–2 h of journey from Johor to downtown Singapore which is only 30 kms away. Commuting by car from Johor is also inconvenient because of Singapore's extremely high road user fees. Meanwhile, Singapore is well developed in terms of public transport within its borders, and the transport system in Johor does not cater to the needs of Singaporeans. This makes the series of transboundary development plans between Johor and Singapore incompatible with its transportation facilities.

To address this issue, the Johor government is replanning its public transportation in terms of improving accessibility, safety and comfort by building more overpasses or underpasses to help people cross the highway; improving public spaces to increase commuters' comfort whilst waiting for public transportation; and adding multiple public routes, aiming to improve the connectivity of cross-border areas and ensure a coordinated development on both sides (Rizzo & Glasson, 2011).

3.2.5 *Spatial Coordination Between Guangzhou and Foshan in China*

In Guangdong Province, China, the 'Guangzhou–Foshan Cooperation' officially started in 2009, aiming to break the administrative barriers between Guangzhou and Foshan for regional integration. In cooperation, the conflict mainly comes from the lack of cost–benefit sharing, reliable guaranteeing and coordinative mechanisms.

In the Heshun interchange and toll station project, its Guangzhou side (West Second Ring Road) was tolled, whilst the Foshan side of the same project (First Ring Road) was free of charge. This conflict of interest resulted in the delay of opening on the Guangzhou side, which led to the postponement of the interchange toll station project. In the Nanzhou water plant project, Shunde (an administrative district in Foshan City), as the water provider, is not allowed to set up industries around the water source and has to pay for the expenses of water protection projects. By contrast, Guangzhou, as the project initiator and the largest beneficiary, did not provide corresponding compensation to Shunde, which led to complaints from Shunde and affected the cooperation at a later stage. The conflict is mainly due to the lack of policy resources and voice in Foshan, and there is no qualified cooperative mechanism. In the Jinshazhou area, both sides plan according to specific events and projects with their own needs, and there are some differences in planning objectives. Given the different planning and approval bodies, the Guangzhou side of the Jinshazhou area has achieved full coverage of the control plan, whilst the Foshan side is only at the planning study stage, resulting in a gap in planning collaboration between the two sides. In addition, the lack of a unified technical dialogue platform also has a technical impact on the collaboration: the coordinated planning system of the two cities is not uniform, and it is not easy to convert. In Guangzhou–Foshan Cooperation, the conflict erupts before it is addressed. In the whole process, from collaboration to implementation, there is no stable, independent department to coordinate and protect the interests of both sides (Chen et al., 2012).

3.3 History and Background of Cooperation Schemes

3.3.1 Development History and Relationship Amongst Guangdong, Macao, Zhuhai and Hengqin

In the 2021 General Plan for Building Guangdong–Macao In-Depth Cooperation Zone in Hengqin, Hengqin is the subordinate of Guangdong Province. Furthermore, Macao is also a provincial level administration. Thus, there are briefly three possibilities for the relationship between Guangdong, Macao and Hengqin:

- Guangdong holds the dominance.
- Macao has dominance.
- Both sides cooperate and share governance.

Before 2021, Guangdong Province was the leading partner in Hengqin cooperation, whilst Zhuhai was the actual implementer and executor and fiscal income in Hengqin was included in Zhuhai's statistics. Therefore, Macao has worried that 'Macao loses the right to speak in developmental issues in Hengqin (Yin, 2020), and the Guangdong side believes that 'Hengqin development must adhere to Guangdong's leadership (Fan, 2009)'. However, in the new stage of cooperation after 2021,

Macao has one more executive deputy director in the top management committee than Guangdong has, which may mean more voices on developmental issues (Central Committee of the Communist Party of China, State Council, 2021). In terms of the tax income distribution, all fiscal income from the cooperation zone before 2024 will go to the government of the cooperation zone. As for the attribution of income afterwards, Macao's Chief Executive said in a press conference: 'Although it is challenging to incorporate GDP into the scope of Macao, Macao also can derive a share of the fiscal income after the establishment of the interest-sharing mechanism in the future' (Ou, 2021). In addition, the central government plays a leading, organising and coordinating role in the Guangdong–Macao cooperation. The Zhuhai Municipal Government, as the actual implementer in previous cooperation, does not have a seat as the chairman/vice director in the Management Committee of the cooperation zone this time but has a position as the relevant head in the subordinate executive committee. In the past Zhuhai–Macao cooperation, there was also a phenomenon that Macao paid more attention to the interaction with the central government and ignored Zhuhai. Therefore, Zhuhai, as the actual executor of the cooperation, has a complex relationship with the Macao government as a non-subordinate and non-equivalent local government, perhaps making it the most complex pair of relations in the cooperation (Yin, 2020).

3.3.2 Analysis of the Structural Function of Planning in Hengqin (As a City in Mainland China)

Urban planning is a crucial tool for the government to manage spatial resources, and land utilisation is the core. Planning can suppress negative impacts of development, enhance positive results and induce harmonious development of social and ecological aspects. China's urban planning operation implements the 'two permissions and one submission (liangzhengyishu)' system, which includes 'Site Submission', 'Planning Permission for Construction Land' and 'Planning Permission for Construction Project'. Such a system has been designed to control the locations of projects, land functions, the density of development, impacts of constructions and so on (Ministry of Human Resources and Social Security in China, 2019). As a developing city in the process of urbanisation, the main objective of planning in Hengqin is to promote economic and urban development by constructing focused areas. This also forms the target for planning Hengqin–Macao cooperation, such as the planning for industrial diversification (Hengqin New District Management Committee, 2016).

3.3.3 Analysis of the Structural Function of Planning in Macao

For Macao, which has primarily completed its urban growth process, planning aims to improve and renew the existing urban space to create a happy, intelligent, sustainable and resilient city. In the past, Macao did not have a macrolevel urban master plan but primarily small-plot implementation designs generated by the city's specific needs. The 'General Statute of Urban Architecture' and related regulations involving land and building management, such as parking and fire regulations, were the primary basis for planning and management. These documents mainly were administrative guidelines within the government. After implementing the 'Town Planning Law' in 2004, this was used as the basis for urban planning, and in 2021, the 'Urban Building Legal System' was published to control the urban construction process. However, it was not until 2022 that Macao's urban master plan was formally completed for the first time.

3.4 Characteristics of Two Systems and Cooperation Schemes

3.4.1 Legal System of Planning

The current system of urban planning regulations in Macao consists of the Legal System of Urban Architecture, the Urban Planning Law and its accompanying administrative regulations, such as the Rules for the Implementation of the Urban Planning Law and the Urban Planning Commission. The Town Planning Law and its accompanying rules prepare, approve, implement, review and revise town planning; they specify the definition, types and levels of planning, the composition of planning documents and the establishment and duties of town planning committees and interdepartmental committees. It also strengthens the implementation of urban planning by providing punishments for violating planning controls'. The Legal System of Urban Architecture is used in the urban construction process, such as construction permits, inspections of construction and punishment for violations (Legal System of Urban Construction, 2021).

Urban planning in Hengqin depends on a more comprehensive and complex legal system than in Macao. The existing planning system is based on a dozens of regulations from the central government, Guangdong Province, Zhuhai City and Hengqin In-Depth Cooperation Zone, ranging from the Law of the People's Republic of China on Urban and Rural Planning to the General Plan for Building Guangdong–Macao In-Depth Cooperation Zone in Hengqin. For example, in 2016, the Detailed Plan of Hengqin New Area is relevant to up to twenty regulations. More regulations mean more constraints. The future cooperation zone, as an area under the direct supervision of Guangdong Province rather than under Zhuhai City, will be subject to fewer restrictions. Nonetheless, it may run in conflict with the Zhuhai city planning regulations system.

The cooperation between Hengqin and Macao is grounded in two completely independent regulatory systems, which inevitably lead to differences and overlaps in the content and binding force of the regulations, which then lead to conflicts and problems in practice, such as which law to follow, who to judge and enforce the planning of the zone, how to punish violations of the law and other aspects of practice based on which regulation, will be applied the two sides. Besides, when the same issue is required differently between the two systems, coordination is a challenge for the in-depth cooperation zone. For example, the Macao side believes that there is an excellent risk of directly terminating the existing urban planning according to Article 26 of the Urban Planning Law. By contrast, the Guangdong side does not have relevant regulations.

3.4.2 System of Plan Formulation

Macao's urban planning is divided into two parts: a master plan and detailed plans. The process of plan formulation is divided into six parts: the Chief Executive's approval to start the formulation process; the preparation of the draft plan by the Land and Urban Construction Bureau, this part should gather opinions from an interdepartmental committee; a 60-day public consultation for the master plan and 100-day public consultations for detailed plans and the collection of the opinions of the powerful stakeholders and the public on the draft plan; publication of the analytic report of public opinions within 180 days after finishing consultation; the collection of the opinion from the Urban Planning Commission (within 60 days); submitting the final report to the Chief Executive, who will make the final decision. After the draft is finalised, it is approved by the administrative regulations published in the Official Bulletin of the Macao SAR. If significant changes are required, then public consultation should be conducted again after the changes are made. During the formulation of the detailed draft plan, it is necessary to collect the opinions and suggestions from stakeholders of private land and the grantees of state-owned land who may be affected by the implementation of detailed plans. In such a case, the Land and Urban Construction Bureau will hold a briefing session. In addition, the Urban Planning Commission has established an auditing system for some planning-related meetings, allowing city residents to observe the meetings upon application (Government Printing Bureau, 2014b; Urban Planning Commission, 2014).

Hengqin's urban planning is divided into two parts: a master plan and detailed control plans. The planning process consists of the following five sections: the preliminary preparation stage, including the groundwork of the project and the collection of basic information; the preparing stage requires consultation with relevant departments and expert meetings; the stage of reviewing by the planning committee, includes steps, such as the public announcement of the draft and the normative review of the planning committee; the successful submission and approval stage, which embraces the announcement of the post-approval results after the submission of documents to the higher government and inputting approval results; in the final stage, the approved files will be issued to all relevant departments, and the results will be archived whilst the quality assessment of the results is conducted (Government Printing Bureau, 2012; Government Printing Bureau, 2014b; Urban Planning Commission, 2014). As for public participation, the previous practice includes pre-approval consultation and post-approval publication, but there are unclear objects, times, formulas, population types and consultation locations.

Planning systems in Hengqin and Macao both contain a master plan and detailed control plans and have departmental synergy steps. However, there is a significant difference in the planning approval process, as the final decision on the preparation of the plan for the Macao side is made by the Chief Executive, whilst the plan for Hengqin needs to be submitted to a higher-level government for approval. Additionally, the discrepancy in public participation is obvious. Explanation of details about public participation in Hengqin during its planning process is obscure, whilst Macao provides clear and specific regulations of public engagement. In addition, the presence of a large number of properties purchased by Macao residents and some communities created specifically for Macao residents (e.g. Macao New Neighbourhood) in the Hengqin In-Depth Cooperation Zone will raise other problems in the future, including which group is the main target of public consultation when different stakeholders have a conflict. Such a case becomes more complicated for Macao residents who do not live in Hengqin but own Hengqin properties, Macao residents who live in Hengqin, transboundary commuters who rent in Hengqin or Hengqin indigenous residents.

3.4.3 System of Planning Management

The Land and Urban Construction Bureau is the core coordinator of planning matters in the urban planning process of Macao. The Interdepartmental Committee coordinates the relationship between the various departments and the various plans (the committee includes the Cultural Affairs Bureau, Home Affairs Bureau, Transport Bureau, Construction Development Office, Transport Infrastructure Office, Environmental Protection Bureau, Housing Bureau, Tourism Bureau) and has the relevant administrative and legal support. The Land and Urban Construction Bureau is responsible for providing technical and administrative support to the Intersectoral Committee and for covering the related financial expenses. The Urban Planning Committee coordinates the public interest; such a commission is composed of representatives of the public administration, professionals in the field of urban planning and other areas related to it and persons recognised as outstanding in communities. In this Committee, there is a chairman and up to 34 members. Within these members, there will be up to 27 distinguished persons from communities. These two committees have the power to make recommendations on the plan, and the Land and Urban Construction Bureau is required to prepare and submit a final report to the Chief Executive within ninety days of receiving the Committee's comments, who will make the final decision. This formal and public system of multi-departmental coordination means that the synergistic departments also bind the planning body in the planning process (Government Printing Bureau, 2012; Government Printing Bureau, 2014b; Urban Planning Commission, 2014).

Hengqin's current participating departments are mainly the Management Committee and the Executive Committee of Guangdong–Macao In-Depth Cooperation Zone in Hengqin. The Urban planning and Construction Bureau is the leader in urban planning and design and is mainly responsible for the preparation and implementation of master plans, detailed control planning, special planning, urban design, land resource management, construction project management, administrative approval and services, including other planning-related content in the cooperation zone (Guangdong–Macao In-Depth Cooperation Zone in Hengqin, 2021). In terms of sectoral collaboration, it is common practice to conduct horizontal consultation with relevant sectors during the plan-formulation phase, which often only serves as suggestions and does not create solid constraints and changes to the planning process.

The differences between the two planning management systems are mainly reflected in the depth of involvement of the relevant coordinating departments in the planning and in who are the final decision makers in the planning. The level of participation of departments in the interdepartmental committee in Macao is relatively higher than in the participation model of Hengqin. Whilst the planning participation department under the Macao model acts as a degree of constraint on the Land and Urban Construction Bureau in the planning preparation process, the planning participation department under the Hengqin model does not have a solid binding effect on the Urban Planning and Construction Bureau. In future planning, when planning-related departments object to the plan, actors in Macao and Hengqin systems may give different importance to the objections, which may lead to certain conflicts. In addition, the Chief Executive has decision-making power in the planning management system of Macao. Although the in-depth cooperation zone has been promoted to be directly under Guangdong Province, planning approval still needs to be submitted to the higher authorities for approval.

3.4.4 System of Planning Practice

The responsible department for the approval of registered urban planners in Macao is the Professional Council of Architecture, Engineering and Urban Planning. The requirements to apply for registration are a person who has obtained a bachelor's degree or higher in urban planning, completed an internship (two years of full-time internship and five years of part-time internship) and is a resident of the Macao SAR; by contrast, a public administration staff member who has obtained a bachelor's degree in urban planning and has worked for three consecutive years in the field of urban architecture or urban planning. After passing the examination, they can become registered urban planners (Legislation for Urban Architecture and Town Planning in Macao Special Administrative Region, 2015). In addition, Macao has no qualification classification for urban planning preparation units as restrictions in operating professional planning issues.

Hengqin complies with the policy of Mainland China that the responsible department for the approval of registered urban and rural planners is led by the Ministry of Housing and Urban–Rural Development and Ministry of Human Resources and Social Security, and the local Municipal Human Resources and Social Security Bureau and Planning Bureau. The requirements to apply for registration are to obtain a degree in urban and rural planning or architecture, engage in urban and rural planning for a certain number of years and pass the professional qualification examination for registered urban and rural planners. Amongst them, a specialist degree requires six years of practice; a bachelor's degree requires three years of practice (three years to attend an institution that has passed professional assessment); a master's degree requires two years of practice, and the number of years of practice in other professions increases by one year according to the academic requirements (Ministry of Human Resources and Social Security in China, 2019). From the perspective of urban and rural planning preparation units, urban and rural planning preparation units in Mainland China are divided into classes A (jiaji), B (yiji) and C (binji). Class A urban and rural planning preparation units are not restricted in their scope of undertaking urban and rural planning preparation business. At the same time, the other levels are subject to corresponding restrictions.

Regarding the planning practice system, the relevant responsible authorities for the registration of planning practitioners are different between Hengqin and Macao, and the requirements for practitioners to apply for registration are also different between the two sides, with Mainland China's requirements being more stringent in comparison. Whether the planning hierarchy in Mainland China applies to the Macao planning system is also an issue in collaborative planning. In 2012, Guangdong and Macao started the interchange of registered architecture, urban planning and landscape architecture design services. However, the relevant bill emphasises that the Comprehensive Urban Plan is not included (Government Printing Bureau, 2012).

3.5 Differences Between the Masterplans of Hengqin and Macao

Macao decided to prepare a draft city master plan in 2018 promulgated in 2011. The process of preparation includes the city planning relevant officials, the city planning coordination meeting (composed of representatives of government departments), the city planning committee (composed of experts, scholars, public representatives and government representatives), the divisional authorities in Macao government (Land and Urban Construction Bureau, 2010). The latest plan, the Developmental Master

Plan of Hengqin New Area, was promulgated in 2021, and the preparation process is the same as that of other cities in Mainland China. The main bodies involved in formulating the plan are Guangdong–Macao In-Depth Cooperation Zone in Hengqin Management Committee and Executive Committee, which are leading urban planning and design. Another important body is the Urban Planning and Construction Bureau, which is mainly responsible for the preparation and implementation of spatial planning, urban design, land resource management, construction project management, administrative approval and services, and other planning-related contents in the cooperation zone.

In particular, in their different masterplans, Macao is committed to building a world-class tourism and leisure centre. At the same time, Hengqin is a new platform to promote the moderate and diversified development of Macao's economy, and there is a massive difference in the general positioning of the two. Regarding disaster prevention, Macao's master plan mentions the design of infrastructure with disaster prevention elements to enhance its response capability. At the same time, Hengqin proposes to adhere to the policy of 'combining prevention, resistance, avoidance and rescue' and to improve flood storage areas and rainwater utilisation facilities, which are also slightly different in terms of relevance. Macao proposes to build a transportation network that combines a slow-moving system, light rail, public buses and other public transport and optimise parking resources. By contrast, Hengqin proposes in its development master plan to establish a high-standard, modern and comprehensive transportation system led by public transport. In terms of the environment, Macao focuses on green, low-carbon and sustainable development. By contrast, Hengqin focuses on building an ecological city with a circular economy as the primary construction mode.

Although the formulation of Macao's master plan started later than that of Hengqin, and there are many differences in many aspects, such as the general positioning, Macao remains to be a role model for other Chinese cities in terms of public participation. For example, in the Urban Planning Law in Macao, it is clearly stated that the 'Land and Urban Construction Bureau shall establish a mechanism to promote public participation in the preparation of the draft urban plan' and explain its content to the public during the public consultation period, in which a distinction is made between the general public and stakeholders. More importance is attached to the views of stakeholders. These established mechanisms in a public hearing in the meetings and the system in protecting individual interests also provide more guarantees of transparency and fairness in the establishment of the system (Wen, 2014).

3.6 Strategies of Cooperation Planning Systems for Hengqin and Macao

Hengqin and Macao have distinct institutional environments, including different planning systems. To coordinate two planning systems to achieve concerted planning and development, five aspects of cooperation may be helpful.

Firstly, a clear and reasonable route map to future changes should be discussed and established. In such a map, the core question is, who is the leader of future development? Macao or Hengqin (Guangdong Province). The leader will have more institutional features to be reflected in future Hengqin. From a gradual perspective, a three steps route map may be suggested. In the next three to five years, it is the first stage of coordination in which it is better to maintain the existing planning system, a Mainland China one, to support rapid growth, which is quite similar to other Chinese cities in the last few decades. After fast-growing, there are some advantages in the Macao system that may be introduced in the second stage for another three to five years. For instance, public participation is a good choice for Hengqin to learn from Macao in formulating planning. Besides, interdepartmental coordination in Macao is another point that can be learnt by Hengqin. In this stage, a gradual systemic transformation from Mainland China to Macao may occur. The third stage is suggested as a Macao-led development phase. Macao will be the leader in building a more international planning system to support a new economic system opening to the world.

Secondly, joint plans, joint investment and joint construction can be applied as core strategies in developing cross-border areas. High-quality construction needs better coordination in the whole process of development. Joint planning with planners and planning officials from both sides will be the first step to start building new urban spaces.

Thirdly, communication between two planning systems is crucial to building trust and mutual understanding. Such communication should be frequent on multiple levels and require both formal and informal channels. It is communication between persons, information, values and experiences.

Fourthly, the role of the central government should be emphasised to resolve possible conflicts between the two systems. As the central government is superior to both sides, therefore, its authority is helpful in dealing with problems when Macao and Hengqin cannot stop quarrelling and be cooperative themselves.

Finally, planning is only one aspect of cross-border cooperation; therefore, planning needs to coordinate with other aspects of social and economic development. For instance, economic diversity is a crucial target for Macao to involve Hengqin in its economic development. Therefore, urban planning in Hengqin should support economic diversity in terms of providing industrial space, research sites, transport facilities and a qualified environment.

References

Chen, X. L., Liao, Y. T., & Liu, S. L. (2012). Discussion on the implementation of city integration under power constraints—A case study of Guangzhou-Foshan City Integration. *Modern Urban Research, 27*(7), 64–69.

Chuah, C., Ho, B., & Chow, W. (2018). Trans-boundary variations of urban drought vulnerability and its impact on water resource management in Singapore and Johor, Malaysia. *Environmental Research Letters, 13*(7), 074011.

European Commission, Committee on Spatial Development. (1999). *ESDP-European spatial development perspective: Towards balanced and sustainable development of the territory of the European union: Agreed at the informal council of ministers responsible for spatial planning in Potsdam, May 1999*. Office for Official Publications of the European Communities.

European Spatial Planning Observation Network. (2007). *Cross-Border cooperation-cross-thematic study of INTERREG and ESPON activities*. Retrieved June 30, 2022, from https://www.espon.eu/sites/default/files/attachments/Cross-Border_Cooperation_web.pdf

Fan, S. J. (2009). The historical orientation and strategic significance of Hengqin development. *Journal of Zhuhai Municipal Administrative College, Party School of the CPC Zhuhai Municipal Committee, 4*, 49–51. Retrieved June 30, 2022, from https://bo.io.gov.mo/bo/i/2022/07/regadm07_cn.asp

Government Printing Bureau. (2012, October 10). *Ce's Notice no. 48/2012*. Imprensa Official. Retrieved June 30, 2022, from https://bo.io.gov.mo/bo/ii/2012/41/aviso48_cn.asp

Government Printing Bureau. (2014a, February 24). *Rules for the implementation of the urban planning ACT, 5/2014*. Imprensa Official. Retrieved June 30, 2022, https://bo.io.gov.mo/bo/i/2014/08/regadm05_cn.asp

Government Printing Bureau. (2014b, February 24). *Urban planning ACT, 12/2013*. Imprensa Official. Retrieved June 30, 2022, from https://bo.io.gov.mo/bo/i/2014/08/regadm03_cn.asp

Government Printing Bureau. (2014c, February 24). *Urban planning commission 3/2014*. Imprensa Official. Retrieved June 30, 2022, from https://bo.io.gov.mo/bo/i/2014/08/regadm03_cn.asp

Government Printing Bureau. (2015, January 5). *The Legislation for urban architecture and town planning in Macao special administrative region, 1/2015*. Imprensa Official. Retrieved June 30, 2022, from https://bo.io.gov.mo/bo/i/2015/01/lei01_cn.asp

Government Printing Bureau. (2020, August 16). *The legal system of urban construction, 14/2021*. Imprensa Official. Retrieved June 30, 2022, from https://bo.io.gov.mo/bo/i/2021/33/lei14_cn.asp

Government Printing Bureau. (2022, February 24). *Approved the master plan for the cities of the Macao special administrative region (2020–2040), 7/2022 2022*. Imprensa Official. Retrieved June 30, 2022, from https://bo.io.gov.mo/bo/i/2022/07/regadm07_cn.asp

Guangdong-Macao In-Depth Cooperation Zone in Hengqin. (2021). *Bureau of urban planning and construction*. Retrieved June 30, 2022, from http://www.hengqin.gov.cn/macao_zh_hans/hzqgl/zzjg/znjs/content/post_2990085.html

Healey, P., Khakee, A., Motte, A., & Needham, B. (1997). *Making strategic spatial plans: Innovation in Europe*. UCL.

Hengqin New District Management Committee. (2016). *Master urban planning of Henqin new area (2014–2020)*. Zhuhai Government, China.

Herzog, L. A. (2020). Cross-border planning and cooperation. *The US-Mexican border environment: A road map to a sustainable* 139–162.

Nadin, V. (2007). The emergence of the spatial planning approach in England. *Planning Practice and Research, 22*(1), 43–62.

Ou, J. S. W. (2021, September 11). *Macau News: Macao gaming will not move to Hengqin' 2021*. Retrieved June 30, 2022, from https://www.waou.com.mo/2021/09/11/

Rizzo, A., & Glasson, J. (2011). Conceiving transit space in Singapore/Johor: A research agenda for the Strait Transnational Urban Region (STUR). *International Journal of Urban Sustainable Development, 3*(2), 156–167.

The Central Committee of the Communist Party of China, State Council. (2021, September 5). *General plan for the development of Hengqin Guangdong-Macao In-depth Cooperation Zone, State Council Bulletin* (vol. 26). Retrieved 30 June, 2022, from http://www.gov.cn/zhengce/2021-09/05/content_5635547.htm

The Land and Urban Construction Bureau of the Macao Special Administrative Region of the People's Republic of China. (2010, November). *The Land and Urban Construction Bureau, 2010. A study on the Urban planning system of Macao*. Retrieved June 30, 2022, from https://urbanplanning.dsscu.gov.mo/cn/download/Estudo01.pdf

The Ministry of Human Resources and Social Security in China. (2019, December 23). *Notice of the Ministry of Human Resources and Social Security and the Ministry of Housing and Urban-Rural Development on printing and distributing Regulations on the Professional Qualification System for Registered Urban and Rural Planners and Measures for the Implementation of the Professional Qualification Examination for Registered Urban and Rural Planners*. People's Government of Guangdong Province. Retrieved June 30, 2022, from http://www.gd.gov.cn/zwgk/wjk/zcfgk/content/post_2725587.html

Tölle, A. (2013). National planning systems between convergence and incongruity: Implications for cross-border cooperation from the German-polish perspective. *European Planning Studies, 21*(4), 615–630.

Wen, Y. (2014). A review on the development and characteristics of Macao's urban planning system. *City Planning, 38*(S1), 23–30.

Yin, Y. F. (2020). Zhuhai-Macao cooperation to develop inter-governmental relations in Hengqin. *Journal of Macao Polytechnic Institute, 4*, 44–54.

Zhang, L. J. (2011). Implications of EU spatial planning and cohesion policy. *Homeland Resources Intelligence, 11*, 36–43.

Zhou, W. (2004). High expectations and low profile: Urban development in Kuala Lumpur and Singapore. *Journal of Guangxi University of Nationalities, 26*(2), 27–37.

Open Access This chapter is licensed under the terms of the Creative Commons Attribution 4.0 International License (http://creativecommons.org/licenses/by/4.0/), which permits use, sharing, adaptation, distribution and reproduction in any medium or format, as long as you give appropriate credit to the original author(s) and the source, provide a link to the Creative Commons license and indicate if changes were made.

The images or other third party material in this chapter are included in the chapter's Creative Commons license, unless indicated otherwise in a credit line to the material. If material is not included in the chapter's Creative Commons license and your intended use is not permitted by statutory regulation or exceeds the permitted use, you will need to obtain permission directly from the copyright holder.

Chapter 4
Industrial Spatial Synergy Development in Hengqin and Macao

Abstract This chapter describes the close connection between the industrial changes and urban development in Macao and Hengqin. The current land use and industrial distribution status in both places are analysed. The development direction, central trend and density of residential and various industries are analysed in GIS. The industrial structures of both Macao and Hengqin serve urban functions. The industrial density in Macao is higher than that in Hengqin, and the balance of employment and residence in the spatial dimension between the two places is insufficient. In the future, Hengqin will improve the city's supporting infrastructure and strengthen the effectiveness of synergistic industrial development.

Keyword Land use · Industrial development · Hengqin

4.1 Land Use Types in Macao and Hengqin

In 2009, the CPC Central Committee and the State Council decided to develop Hengqin to create an essential platform for promoting the moderate and diversified development of Macao's economy, taking advantage of Hengqin being separated from Macao by a bridge and water. It is a new platform to promote the moderate and diversified economic development of Macao, a new space to improve life, a recent demonstration to enrich the practice of 'One Country, Two Systems' and a new highland to promote the construction of the Guangdong–Hong Kong–Macao GBA (Zheng, 2009). The master plan places a prominent position on the 'development of new industries that promote the moderate diversification of Macao's economy' and specifies the development of four major categories of 'new industries': technological research and development and high-end manufacturing industries, Chinese medicine industry, cultural tourism, exhibition and trade industries, and modern financial industry (Central Committee of the Communist Party of China State Council, 2021). Macao lacks land resources, and its production costs are higher than the mainland. At the same time, the Hengqin Guangdong–Macao Deep Cooperation Zone has enough space to transform scientific and technological achievements into practice.

The land use and land area ratio of Macao and Hengqin is analysed (Figs. 4.1 and 4.2). The proportion of the area of each type of land in Macao is more balanced compared with Hengqin.

Amongst the various types of land use in Macao, the residential area accounts for 23% of the total area, and the industrial area accounts for 2% of the entire area. Ecological protection zones and public infrastructure zones account for approximately 20%, of which ecological protection zones and the open spaces account for approximately 30%, indicating that the Macao government highlights 'ecological

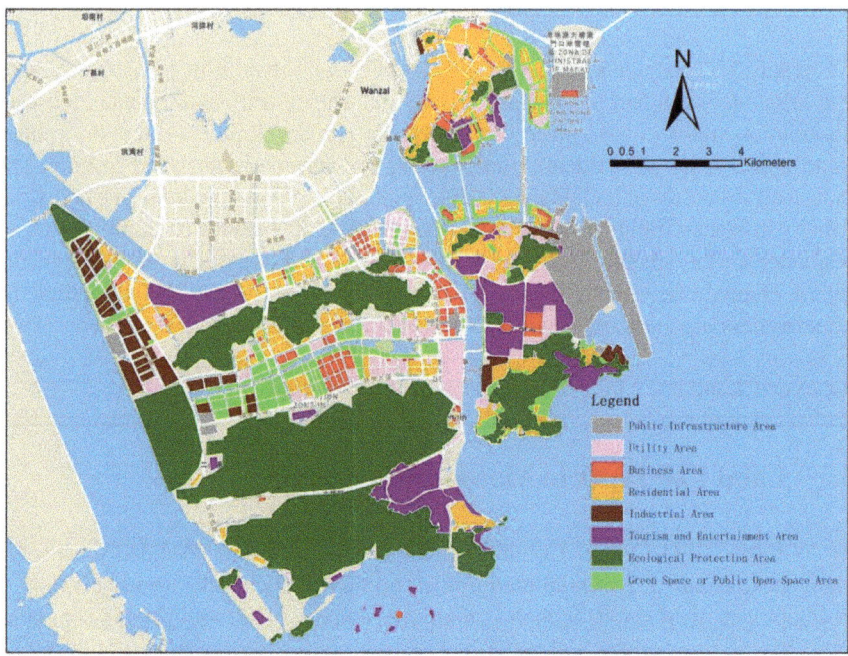

Fig. 4.1 Land use map of Macao and Hengqin

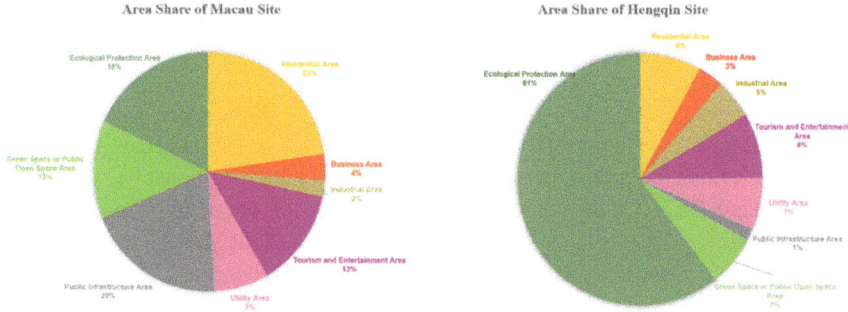

Fig. 4.2 Land proportions in Macao and Hengqin

priority' in the urbanisation process and long-term planning for the sustainable and healthy development of the city. The tourism and recreation area accounts for 13% of the total area, which is an attractive place to provide rest and recreation for tourists in Macao, where land resources are tight, and is in line with the city's positioning as a world-class tourism and recreation centre.

According to the 'Hengqin Master Development Plan', Hengqin has designated four spatial control zones in the development and construction: No Build Zone, Limited Build Zone, Suitable Build Zone and Built Zone, of which the No Build Zone is 57.90 km^2, more than half of the area of Hengqin, strictly protecting the original ecological environment of Hengqin. The minor proportion is the public infrastructure zone, which needs to be increased in future planning and construction.

4.2 Macao and Hengqin Industry and POI Distribution

Living service facilities are the central part of urban public service facilities, and the supporting facilities of urban residential areas also affect the quality of life of urban residents. Moreover, the layout of living service facilities affects urban traffic, resource allocation and residents' quality of life. With the application of big data in spatial analysis, new technical means provide new methods for spatial analysis. Amongst them, point-of-interest (POI) data are a type of point data that can represent real geographic entities and generally contains basic information, such as name, category, latitude and longitude and address of point elements.

The POI data used in this section are obtained via the Internet open platform, which was acquired in 2018. The focus is on the urban residential and non-residential spaces in Macao and Hengqin. Thus, POI data from nine areas, including residences, schools, hospitals, government agencies, supermarkets, food markets, banks, offices and scenic spots (parks), are analysed and discussed. Using GIS spatial analysis methods, such as spatial autocorrelation analysis, standard deviational ellipse, central element analysis and kernel density analysis, spatial pattern distribution and convenience of residential and non-residential points in Macao Peninsula, Cotai and Hengqin, are analysed to provide a scientific basis for the future spatial layout development of residential and non-residential development in the two areas.

4.3 Analyses of the Industrial Structures of Macao and Hengqin

Macao's industrial structure and urban transformation are closely linked, and they interact with each other to promote the city's development. The major industries in Macao are tourism and gambling, banking and finance, real estate and construction and cultural exhibition. Macao is supported by building 'one centre, one platform and one base', namely a world tourism and leisure centre, a service platform for trade and commerce cooperation between China and Portuguese-speaking countries and a base for exchange and collaboration with Chinese culture as the mainstream and multiple cultures coexisting.

The planning of Hengqin's industrial structure has been an essential issue of economic cooperation between Guangdong and Macao since the 1880s. In 1992, the Guangdong Provincial Government identified Hengqin Island as a key development area. In 2021, the Master Plan of Hengqin Guangdong–Macao Deep Cooperation Zone had planned four major industries: science and technology and manufacturing industry, Chinese medicine industry, tourism and exhibition industry and modern financial industry.

The spatial distribution of the four pillar industries in Macao and Hengqin is analysed in GIS. The results are shown in Fig. 4.3. Most of the pillar industries in Macao are distributed in the Macao Peninsula, amongst which the construction and real estate industries are the most widely distributed and the most numerous, followed by the tourism and gaming industries, which are more evenly distributed. Hengqin's pillar industries are mainly modern financial and cultural tourism, exhibition and commerce, concentrated in the northeast of the 'financial island' and Hengqin port. The science and technology industrial park and the Chinese medicine industry are primarily under planning and construction. The number of the four pillar industries in Macao and Hengqin and the ratio of the four pillar industries to other industries are shown in Table 4.1. When comparing the two regions, the number of sectors in Macao is more significant than that in Hengqin, whilst Hengqin still has a shortcoming in terms of quantity. In the future, on the basis of the existing advantages, the two places will cooperate strongly to promote Macao's profitable industries to facilitate Hengqin's high-quality development.

4.4 Diversity Analysis of the Urban Block Spatial Function

Streets play a crucial role in urban life as the primary carrier of transportation and a virtual urban open space. It is not only the main transportation carrier but also a virtual urban open space, which is the basic unit for residents to know the city and urban life (Wang, 2002). The concept of mixed urban land use originates from the mixed primary use proposed by Jacobs (1992), which is widely used but still has ambiguity in its connotation. The American Planning Association believes that diverse functions are appropriate for walking and creating high-density, diverse spaces compatible with different functions. Real estate industry organisations in the U.S. also generally believe that mixed-function is an integrated development that includes residential, retail, office and entertainment functions and seeks to maximise the use of space, and it also includes a combination of different functional types of adjacent parcels in a particular area.

The diversity of spatial functions is generally measured by the formula of information entropy. The mixed degree of function of the research neighbourhood is a

4.4 Diversity Analysis of the Urban Block Spatial Function

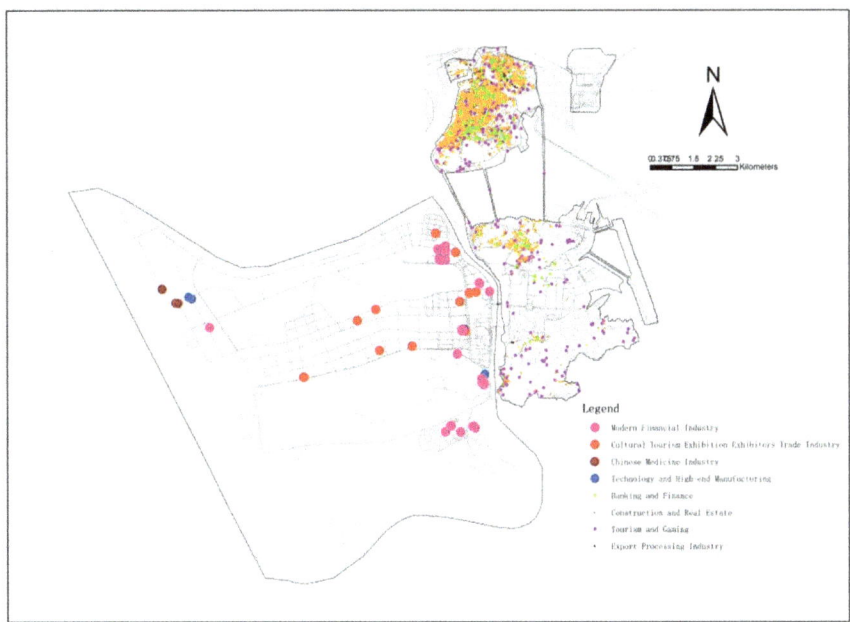

Fig. 4.3 Distribution of pillar industries in Macao and Hengqin

Table 4.1 Structure of the four pillar industries in Macao and Hengqin

Area	Industry	Number	Percentage of other industries, POI (%)
Macao	Export Processing Industry	140	0.50
	Tourism and Gaming	459	1.90
	Construction and Real Estate	3708	15.10
	Banking and Finance	186	0.80
	Total	4493	18.40
Hengqin	Technology and High-End manufacturing	6	1.00
	Chinese Medicine Industry	3	0.50
	Cultural Tourism Exhibition Exhibitors Trade Industry	12	2.00
	Modern Financial Industry	52	8.80
	Total	73	12.40

diverse method of meeting the different needs of the population, and the people can interact, communicate and implement a variety of multiform activities in such spaces. Combined with the existing results measurement method, according to the research object space reasonable adaptation (Long & Zhou, 2016), the formula for measuring the degree of functional mixture of neighbourhood space is set as follows:

$$I = \sum_{i=1}^{n}(p_i * \ln(p_i)), \tag{4.1}$$

where I represents the calculation target of block function mixing degree; i represents the number of interest points in the block i unit; and P_i represents the ratio of interest points in the block I unit to the total number of interest points in the range. Entropy is used to characterise the degree of functional mixing at the street scale.

The MOLAIN index (Moran's I) is a commonly used measure of the spatial clustering of mixed functions in a block (Du, 2014). The formula is set as follows:

$$MI = \frac{n\sum\sum w_{ij}(x_i - \overline{x})(x_j - \overline{x})}{\sum\sum w_{ij}(x_i - \overline{x})^2}, \tag{4.2}$$

where X_i represents the attribute value of region i; n represents the total number of regions; and W represents the adjacent space weight matrix in the binary system, and the rule is expressed as

$$w_{ij} = \begin{cases} 1 \\ 0 \end{cases}$$

where 1 represents regions i and j contiguous; 0 represents regions i and j not contiguous. It is generally assumed that the MOLAIN index statistic follows a normal distribution with an expected value. When the calculated value of the MOLAIN index is less than, it is assumed that there is a negative spatial correlation, and vice versa, it is assumed that there is a clustering pattern.

As shown in Figs. 4.4, 4.5 and 4.6, there is a positive spatial correlation between Macao and Hengqin, and the functional mix and the spatial autocorrelation in Macao neighbourhoods are higher than in Hengqin communities. The neighbourhood mix is generally higher in Toi San, Areia Preta, Iao Hon, and Praia Grande e Penha have higher mixed degree, which can help to meet the daily needs of residents well. The mixed degree of the NAPE e Aterros da Baía da Praia Grande is lower, not conducive to crowd communication or activities. The mixed degree of Hengqin district is clearly distributed in a ring shape.

4.5 Analysis of the Directional Distribution and Central Trend of Residential and Non-Residential Points

Standard deviation ellipse analysis is a spatial statistical method for analysing the point data. It is used to measure the direction and distribution range of a set of data. The long half-axis of the ellipse represents the direction of the data distribution

4.5 Analysis of the Directional Distribution and Central Trend ...

Fig. 4.4 Mixing degree of Macao blocks

and the short half-axis represents the range of the data distribution. The larger the difference between the long and short semi-axes, the larger the flatness and the more obvious the directionality of the data (Zhang et al., 2018). In this study, a doubled standard deviational ellipse was used as follows:

$$\tan\theta = \frac{\sum_{i=1}^{n}(x_i-\bar{x})^2 - \sum_{i=1}^{n}(y_i-\bar{y})^2 + \sqrt{\left[\sum_{i=1}^{n}(x_i-\bar{x})^2 - \sum_{i=1}^{n}(y_i-\bar{y})^2\right]^2 + 4\left[\sum_{i=1}^{n}(x_i-\bar{x})^2 - \sum_{i=1}^{n}(y_i-\bar{y})^2\right]^2}}{2\sum_{i=1}^{n}(x_i-\bar{x})\sum_{i=1}^{n}(y_i-\bar{y})}, \quad (4.3)$$

$$\sigma_x = \frac{\sqrt{\left[\sum_{i=1}^{n}(x_i-\bar{x})\cos\theta - (y_i-\bar{y})\sin\theta\right]^2}}{n}, \quad (4.4)$$

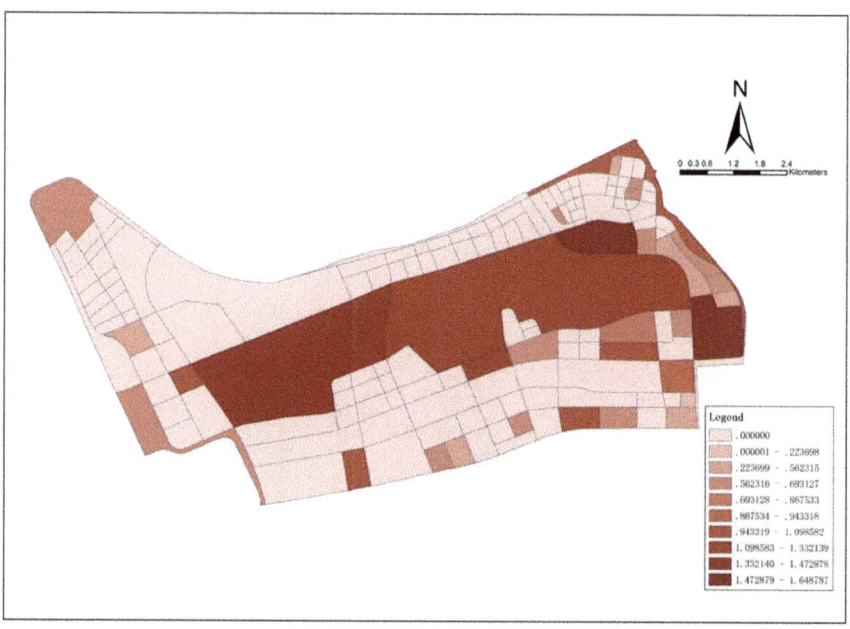

Fig. 4.5 Mixing degree of Hengqin blocks

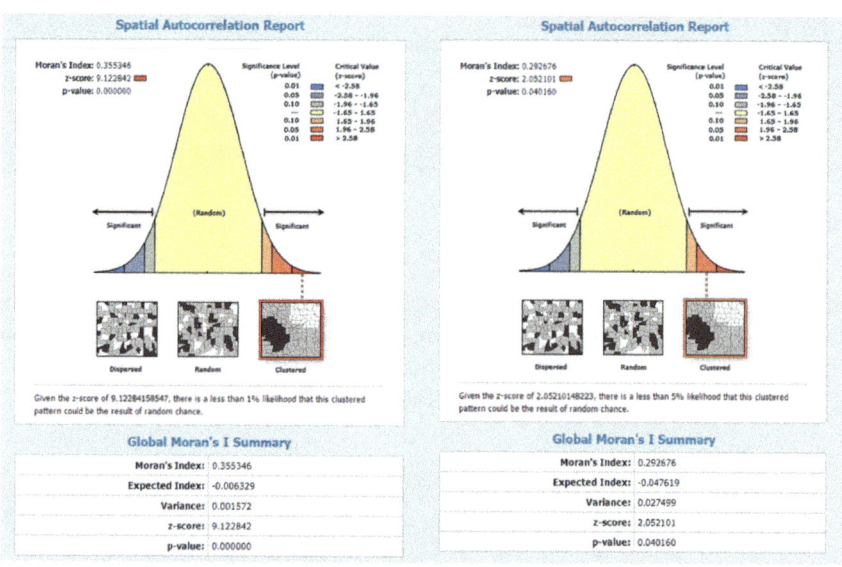

Fig. 4.6 Moran I indices of Macao (0.35) and Hengqin (0.29)

4.5 Analysis of the Directional Distribution and Central Trend … 59

Table 4.2 Classification and number of POIs in Macao and Hengqin

	POI Type	Macao	Hengqin
Residential	Residential	3700	12
Non-Residential	School	150	9
	Hospital	110	30
	Government organisation	760	60
	Supermarket	200	50
	Vegetable market	200	20
	Bank	190	50
	Office Space	560	40
	Scenic Spots (Parks)	459	60
Total		6329	331

$$\sigma_y = \frac{\sqrt{\left[\sum_{i=1}^{n} (x_i - \overline{x}) \sin \theta + (y_i - \overline{y}) \cos \theta\right]^2}}{n}, \quad (4.5)$$

where X_i and Y_i represent the cognitive coordinates of the i location, and n is the number of locations; $\overline{x}$ and $\overline{y}$ denote the average of the *x*-coordinate and *y*-coordinate values of all points; and θ denotes the rotation direction angle.

The POI data used in this study were obtained from online open Maps. The data were de-duplicated, merged and reclassified. Finally, 6660 valid data were retained. The data were initially divided into the residential and non-residential categories. Then the non-residential points were classified into nine categories: residences, schools, hospitals, government agencies, supermarkets, food markets, banks, office buildings and scenic spots (parks) (Table 4.2).

4.5.1 Residential Standard Deviational Ellipse with Central Element Analysis

The standard deviational ellipse analysis was conducted for the residential points in Macao Peninsula, Cotai and Hengqin (Fig. 4.7). The central element of the residential space of Macao Peninsula was located in Estrada de Coelho do Amaral. The central elements of the residential space of Cotai were located in Avenida de Guimarães with an ellipse area of 7,442,000 m². The central elements of the residential space of Hengqin were located in Avenida de Hong Kong and Macao with an ellipse area of 39,718,000 m². Figure 4.8 shows the spatial distribution analysis after taking consideration of the urban integration in the Macao Peninsula and Hengqin. The standard deviational ellipse of residential space is not significantly offset from the

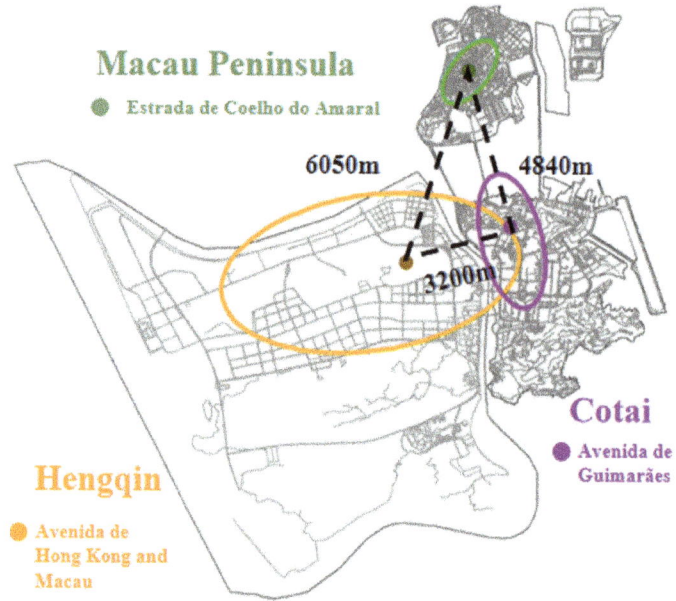

Fig. 4.7 Standard deviational ellipses of residential points in Macao Peninsula, Cotai and Hengqin

Macao Peninsula. After integrating Cotai and Hengqin residential points, the results are shown in Fig. 4.9. The offset is more significant, with a distance of 470 m covering the area near the Hengqin port.

After the urban integration of residential points in Macao and Hengqin, the shift of residential concentration in space is not obvious. This finding could be explained by the current situation that Macao has high population density, dense residential space and scarce land resources. Meanwhile, Hengqin has low population density, scattered residential space and abundant land resources. In the future, it is recommended to create a more convenient living environment for their daily life, promote the planning of urban public facilities, improve the sustainable development momentum of the city and attract more local Macao residents to live in Hengqin.

4.5.2 Non-Residential Standard Deviational Ellipses with Central Element Analysis

The results of the standard deviational ellipse analysis for the non-residential points of Macao Peninsula, Cotai and Hengqin are shown in Fig. 4.10. The central element of the non-residential space of the Macao Peninsula was located at Estrada do Repouso, with an elliptical area of 3936,000 m^2. The central element of the residential space of Cotai was located at Estrada da Baía de Nossa Senhora da Esperança with an oval

4.5 Analysis of the Directional Distribution and Central Trend … 61

Fig. 4.8 Standard deviational ellipses of integration residential points in Macao Peninsula and Hengqin

Fig. 4.9 Standard deviational ellipses of integration residential points in Cotai and Hengqin

area of 11,140,000 m². The central element of the residential space of Hengqin was located at Hengqin Avenue, with an elliptical region of 33,884,000 m². After the non-residential points of Macao Peninsula and Hengqin are integrated, the results are shown in Fig. 4.11. The standard deviational ellipse of non-residential space was not significantly shifted to the southwest with a distance of 300 m. The standard deviational ellipse covers the 'financial island' area north-east of Hengqin. The results after urban integration of non-residential points in Cotai and Hengqin are shown in Fig. 4.12. The spatial discrepancy is 1570 m, and the standard deviational ellipse covers the dense residential area in the central part of Hengqin.

After integrating non-residential points in Macao and Hengqin, the industrial concentration point shifts in space. To a certain extent, it indicates that the change in industrial layout can drive the development of industries in Hengqin and Macao. The industrial structure and POI of both Macao and Hengqin serve the urban functions. The density of POI in Macao is higher than that in Hengqin, and the spatial dimension of the two places is insufficient to balance the jobs and housing. It is suggested that Hengqin needs to improve supporting infrastructures and transportation services to consolidate the existing advantageous resources, develop high-tech industries and enhance the core competitiveness of the region.

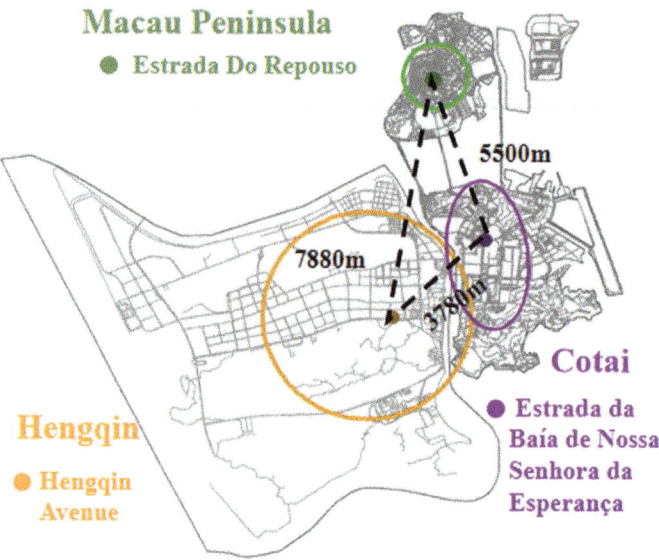

Fig. 4.10 Standard deviational ellipses of non-residential points in Macao Peninsula, Cotai and Hengqin

4.5 Analysis of the Directional Distribution and Central Trend … 63

Fig. 4.11 Standard deviational ellipses of non-residential points in Macao Peninsula and Hengqin

Fig. 4.12 Standard deviational ellipses of non-residential points in Cotai and Hengqin

4.6 Density Analysis of Residential and Non-Residential Points

The distribution pattern and density of POI points in urban space are essential in infrastructure planning and urban spatial analysis. The kernel density estimation method to express this feature is based on the two-dimensional extended Euclidean spatial theoretical system and considers the locational influence of the First Law of Geography. It assigns different weights to the points in the region, and the points close to the centre are given larger weights and vice versa as following:

$$\int h(x) = \frac{1}{nh^d} \sum_{i=}^{n} K\left(\frac{x - x_i}{h}\right), \qquad (4.6)$$

where $k\left(\frac{x-x_i}{h}\right)$ is a kernel function, with a symmetric single-peak probability density function; h is a free parameter that defines the magnitude of the smoothing, d is the dimension of the data; and n is the number of points in the bandwidth range. To understand the geographical concentration of residential space and non-residential space in Macao and Hengqin more directly, a series of visual comparisons between residential space and non-residential space are visualised in Figs. 4.13, 4.14 and 4.15.

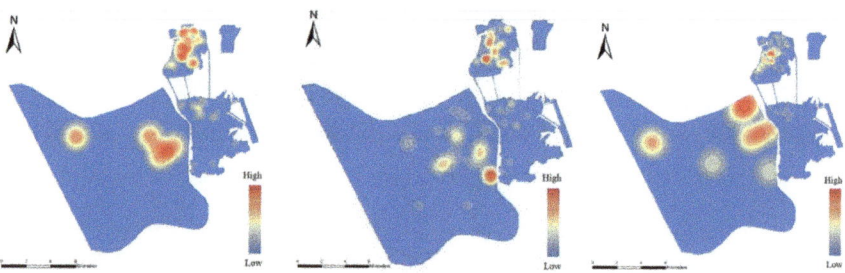

Fig. 4.13 Analysis of the kernel densities of residential areas, schools and hospitals in Macao

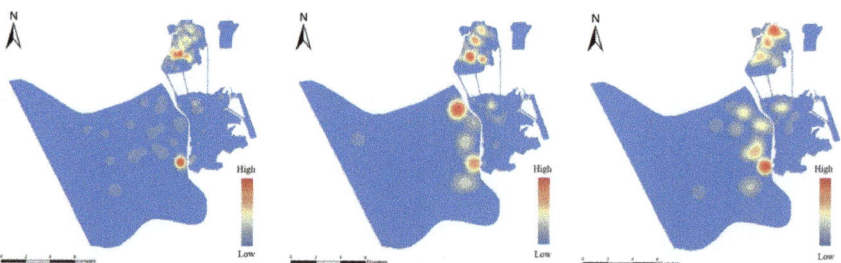

Fig. 4.14 Analysis of the kernel densities of government agencies, banks and office buildings in Macao

4.6 Density Analysis of Residential and Non-Residential Points

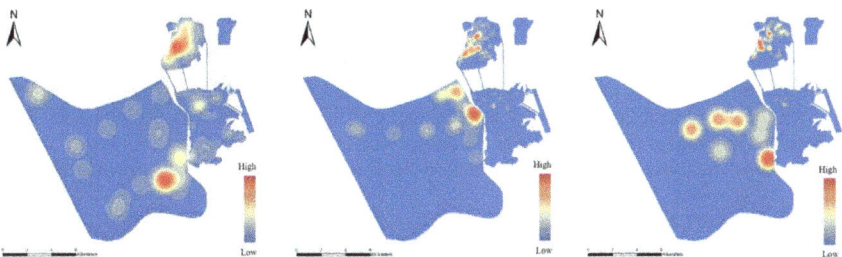

Fig. 4.15 Analysis of the kernel densities of scenic spots (parks), supermarkets and vegetable market in Macao

In general, there are polycentric distribution characteristics of residential points in Macao and Hengqin. Macao Peninsula is more clustered and widely distributed, whilst Cotai is mainly clustered in Taipa Old Town. Hengqin is mainly clustered in the area near Hengqin Port and the north-east. Compared with Hengqin, the overall non-residential spatial distribution in Macao is more affluent, with a more extensive spatial distribution, a higher degree of aggregation and more extensive population coverage. In particular, the high-density concentration areas in the Macao Peninsula have high population densities and public services, such as hospitals, schools, supermarkets, etc., The high-density concentration area of Hengqin is mainly located in Hengqin Port, the northeastern 'financial island' and Zhuhai Changlong, where banks, schools, office buildings, scenic spots (parks) and other facilities are clustered. It is consistent with the functional layout of 'three areas and ten districts' in the Hengqin Master Development Plan.

With the continuous construction and development of the Hengqin New Area, Macao is suggested to fully utilise the advantages and strengthen the complementary development of industries in both places, thus upgrading and optimising the regional industrial structure. The advantages of Macao's world-class leisure and tourism and its economic and trade platform can be used to attract enterprises and talents to Macao whilst focusing on the cultivation of talents and strengthening the industrial integration between the two places. Given there are differences in the industrial structure and resource endowment between Macao and Hengqin, reasonable planning is needed for the effective allocation of resources to prioritise the development of major industries.

References

Central Committee of the Communist Party of China State Council. (2021). Hengqin Guangdong-Macao deep cooperation zone construction master plan. *Communique of the State Council of the People's Republic of China, 26*, 7–13.

Du, J. (2014). Analysis of the global distribution pattern of theft crime rate based on Moran index. *Police Technology, 01*, 45–46.

Jacobs, J. (1992). *The death and life of great American citie*. Random House.

Long, Y., & Zhou, Y. (2016). Quantitative evaluation of street dynamics and analysis of influencing factors: Take Chengdu as an example. *New Architecture, 01*, 52–57.

Wang, P. (2002). *Systematic construction of urban public space*. Southeast University Press.

Zhang, J., Chen, S., & Mapunda, D. (2018). Evolution and spatial distribution of the urban system in Tanzania. *World Regional Studies, 27*(1), 22–33. https://doi.org/10.3969/j.issn.1004-9479.2018.01.003

Zheng, Q. G. (2009). "One Country, Two Systems", "Pan-PRD" Cooperation and Hengqin Development. *Pi Shu Database*. https://www.pishu.com.cn/skwx_ps/initDatabaseDetail?siteId=14&contentId=7998205&contentType=literature

Open Access This chapter is licensed under the terms of the Creative Commons Attribution 4.0 International License (http://creativecommons.org/licenses/by/4.0/), which permits use, sharing, adaptation, distribution and reproduction in any medium or format, as long as you give appropriate credit to the original author(s) and the source, provide a link to the Creative Commons license and indicate if changes were made.

The images or other third party material in this chapter are included in the chapter's Creative Commons license, unless indicated otherwise in a credit line to the material. If material is not included in the chapter's Creative Commons license and your intended use is not permitted by statutory regulation or exceeds the permitted use, you will need to obtain permission directly from the copyright holder.

Chapter 5
Transportation Integration Development in Hengqin and Macao

Abstract This chapter first analyses the accessibility of public transportation in Macao and Hengqin based on the plans in the draft 'Macao Land Transport Master Plan (2021–2030)'. Secondly, the accessibility of schools, bus stops, and other public services in Macao and Hengqin is quantitatively assessed by applying the two-step floating catchment area (2SFCA) method to analyse the adequacy of public services in terms of population density in the transportation areas. Finally, the equity of accessibility in the two areas is compared using the Gini coefficient.

Keyword Accessibility · Public transportation · Hengqin · Macao

5.1 Public Transport Systems in Macao and Hengqin

As a means for assessing the accessibility of urban public transportation, accessibility plays a vital role in the layout of urban networks, urban land use and the construction of urban public transportation systems. An effective evaluation of the accessibility of urban public transportation services is important to optimise the allocation of urban public transportation resources and to meet the public transportation needs of residents better. In 2011, the Macao SAR Government announced the 'Overall Land Transport Policy for Macao (2010–2020)'. Over the past decade, with the gradual promotion and implementation of various action plans and measures in the policy, Macao's land transport environment has been improved evidently. Nonetheless, amidst rapid socioeconomic development, there are also issues and new challenges that require continuous follow-up. Currently, Macao is integrating into the national development situation and facing significant development opportunities, such as Guangdong–Hong Kong–Macao GBA and Guangdong–Macao In-Depth Cooperation Zone in Hengqin. Against this background, the Macao SAR Government has launched the 'Macao Land Transport Master Plan (2021–2030)'. In the plan, by 2030, the public transport share (proportion of light rail, bus and cab trips to total motorised trips for residents and tourists) will increase from 55 to 60%, and the green travel share (proportion of public transport and slow-moving transport to total trips) will reach as high as 75%. By accessing several high-speed railroads and intercity railroads and the introduction of the Macao Light Rail Line, Hengqin will form a 'multi-network' gateway transportation hub, with convenient interchange between various transportation modes, providing more convenient transportation services for the Hengqin Guangdong–Macao Deep Cooperation Zone and the future development of Macao.

5.2 Accessibility to Public Transport in Macao and Hengqin

Network analysis is a tool module for modelling the construction of geographic and urban infrastructure networks, mainly for resource optimisation and allocation network structure improvement. The method is based on road networks and simulates the way people travel in actual conditions under specific resistance. The primary network generally consists of centres, connections and nodes (Zhang et al., 2022). Following the master plan for the city of the Macao SAR (2020–2040), the land of Macao was rezoned into 18 planning areas, considering the population distribution, characteristics and functions of the current statistical areas, as shown in Fig. 5.1. This section analyses the accessibility of public services in Macao and Hengqin based on the coverage of light rail and bus stations in Macao and Hengqin in the spatial scale of residential blocks.

As shown in Figs. 5.2 and 5.3, compared with the Macao light rail, the Hengqin light rail needs to further raise its service population and coverage areas. Macao light rail can serve 26% of the population and 21% of the land area within the service range of 500 m. If the service range is set as 1000 m, then the coverage area and population will reach over 50% (Table 5.1). If the service range of Hengqin light rail is set at 500 or 1000 m, then the coverage population remains unchanged at approximately 24%, and the coverage area increases from 0.7% to 2% (Table 5.2).

There are only three stations of Hengqin rail transit at present, which have not formed a ring or network line, and it is difficult to take full advantage of the rail transit. In the future, with the increase in population and the improvement of the rail transit system, the light rail stations will have a positive guiding effect on the city's development.

Table 5.3 shows the types of POI included in the 500- and 1000-m light rail stations in Macao and Hengqin. The number or types of industries covered by the light rail stations in Macao are larger than those in Hengqin. Within the 500-m light rail coverage of Macao, shopping and catering services are the most popular, followed by transportation facilities and life services. Within the 1000-m light rail coverage of Macao, the number of other types of industries increase significantly.

The distribution of industries is relatively small in the 500- and 1000-m buffer zones of the Hengqin Intercity Railway. Within 500 m, only a small amount of accommodation services, business residences and transportation facilities are distributed to satisfy the residents' needs. Nonetheless, other supporting facilities

Fig. 5.1 Zone planning of Macao

need further improvement. The construction of supporting facilities needs to be further strengthened (Table 5.4).

Figure 5.4 and Table 5.5 show the population and area covered by Macao bus stations convenient service. The 500-m buffer zone covers all construction land in Macao, covering 97.5% of trehe population. Compared with the distribution in Macao, bus stations are somewhat sparsely distributed in Hengqin. The population coverage rates of 300 and 500 m are not different evidently, which are 20.64 and

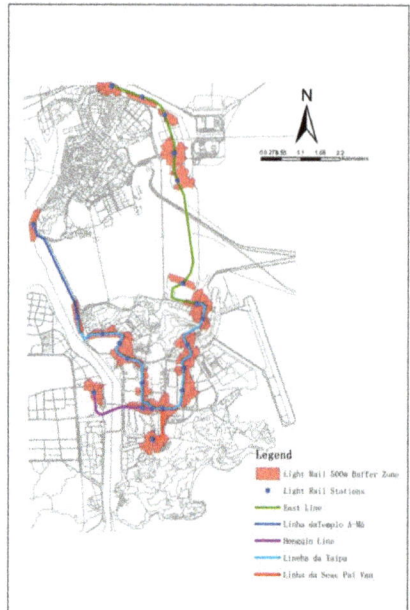 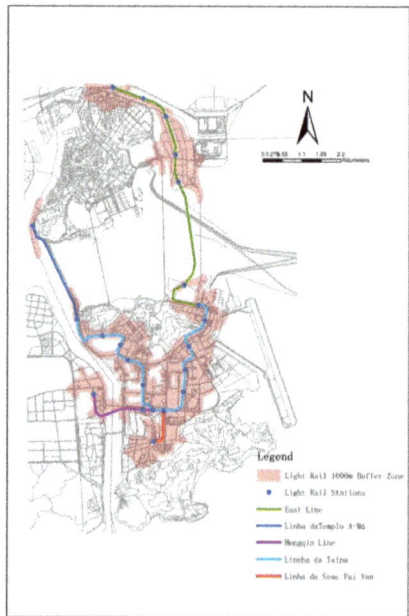

Fig. 5.2 500- and 1000-m service areas of planned light rail stations in Macao

27.62% (Fig. 5.5 and Table 5.6). The coverage area is relatively small, accounting for 6.3 and 11.2% of the total area.

5.3 Accessibility to Public Services in Macao and Hengqin

Public facilities provide public service for the public, including education, medical and health, culture, transportation and other service facilities. Their reasonable layout is directly related to the fairness and efficiency of the allocation of government public service resources, the quantity and quality of public services enjoyed by the public and the goal of equalisation of basic public services.

The 2SFCA method is an essential method in the study of spatial accessibility of public service facilities. It has been widely used in the study of the layout of public service facilities at home and abroad. On the basis of the supply and demand locations, the two-step mobile search method compares the number of resources or facilities accessible to residents within the search radius (Xing et al., 2020).

Taking the accessibility of schools as an example, it is measured by considering the supply–demand relationship of accessibility within a certain range of schools at the supply point and residential areas at the demand point, as shown in Fig. 5.6. Calculate the supply ratio R_j for all facilities J:

5.3 Accessibility to Public Services in Macao and Hengqin

Fig. 5.3 500- and 1000-m service areas of the Hengqin Intercity Railway

$$R_j = \frac{R_j}{\sum_{k \in (d_{kj} \leq d_j)} f(d_{kj}) P_k}, \tag{5.1}$$

where S_j is the total supply of the facility; P_K is the number of search populations; and d_{kj} is the actual distance from each community to the service facility, obtained by calculating the O-D cost matrix in GIS. Calculate the demand point i as the centre, search for all supply points j within the radius, and then sum up the supply and

Table 5.1 Light rail stations coverage in Macao

District name	Covered population (%)		Covered area (%)	
	500 m	1000 m	500 m	1000 m
UOPG Central–3	1.5	3	2.7	4.1
UOPG Este–1	62	2	2	2
UOPG Este–2	8.7	14	2.7	4.1
UOPG Norte–2	2	2	2	2
UOPG Norte da Taipa–2	0.5	1.7	3	10.2
UOPG Taipa Central–2	2	2	2	2
UOPG Taipa Central–1	0.3	7	2	4.1
UOPG Pac On	2	2	2	2
UOPG Cotai	1.4	4.3	5.9	17.8
Total	26	55	21	40

Table 5.2 Light rail station coverage in Hengqin

Buffer area (m)	Covered population		Covered area	
	Quantity	Percentage	Quantity	Percentage
500	17,331	20	0.79	0.7
1000	20,619	24	2.21	2.0

demand ratio of each supply point to obtain the feasibility value:

$$A_j = \sum_{j \in (d_{ij} \leq d_i)} \int (d_{ij}) R_j \quad (5.2)$$

where A_j is the reachability value; d_{ij} is the actual distance from each service facility to each district; di denotes the search radius centred on the demand point.

According to the 2SFCA method, the accessibility of Macao schools to residential areas was obtained (Fig. 5.7 and Table 5.7). Overall, the accessibility of schools in Macao is distributed in a face-to-face manner. The highest accessibility area is Patane e São Paulo. Notably, the accessibility of the ZAPE is also relatively high, probably due to the relatively low population density.

The accessibility of bus stations in Macao is characterised by 'two sides and many points' (Fig. 5.8 and Table 5.8). The 'two sides' are the ZAPE and the NAPE e Aterros da Baía da Praia Grande and Patane e São Paulo. The accessibility in these two areas is relatively high. The 'multiple points' are mainly in the Fai Chi Kei, Móng Há e Reservatório, Barra and Barca, distributed in a dotted pattern.

In addition to the education and public transport spots, the accessibility to other public service facilities in Macao is shown in the left figure, showing a trend of 'high in the South and low in the North', which is also associated with the population

5.3 Accessibility to Public Services in Macao and Hengqin

Table 5.3 Types of POIs included in the 500- and 1000-m light rail stations in Macao

Macao POI point category	500-m buffer zone		1000-m buffer zone	
	Quantity	Percentage (%)	Quantity	Percentage (%)
Catering service	226	0.78	1107	3.84
Scenic spots	12	0.04	64	0.22
Public facilities	10	0.03	25	0.09
Company enterprise	39	0.14	301	1.04
Shopping services	234	0.81	1111	3.85
Transportation services	136	0.47	402	1.39
Financial insurance services	18	0.06	77	0.27
Science, education and culture services	12	0.04	91	0.32
Auto repair services	7	0.02	96	0.33
Business residence	71	0.25	347	1.20
Life services	122	0.42	470	1.63
Sports and leisure services	55	0.19	137	0.48
Healthcare services	81	0.28	323	1.12
Government agencies and social organisations	21	0.07	91	0.32
Accommodation services	7	0.02	43	0.15
Total	1051	3.70	4685	16.30

density distribution in Macao (Fig. 5.9 and Table 5.9). Areas with low population density usually have better accessibility and relatively sufficient resources per capita.

The accessibility to schools, public transport and other public services in the main research area of Hengqin is also analysed in GIS. The accessibility to school in Hengqin is generally low. The areas with the highest accessibility are K3 South Litchi Bay District and Huafa Capital (Fig. 5.10 and Table 5.10). Hengqin senior high school is currently under construction. Therefore, this study does not include senior high school and UM.

The accessibility to bus stations in Hengqin is distributed in a 'single point' near Yinxin Garden and the border port (Fig. 5.11 and Table 5.11). The accessibility to other public services in Hengqin is similar to the distribution of schools, with better accessibility around several major residential areas. However, compared with Macao, accessibility still needs to be improved substantially (Fig. 5.12 and Table 5.12).

Table 5.4 Types of POI included in the 500- and 1000-m light rail stations in Hengqin

Hengqin POI point category	500-m buffer zone		1000-m buffer zone	
	Quantity	Percentage (%)	Quantity	Percentage (%)
Catering service	0	0.00	0	0.00
Scenic spots	0	0.00	0	0.00
Public facilities	0	0.00	0	0.00
Company enterprise	0	0.00	0	0.00
Shopping services	1	0.16	9	1.40
Transportation services	4	0.62	5	0.78
Financial insurance services	0	0.00	0	0.00
Science, education and culture services	0	0.00	0	0.00
Auto repair services	0	0.00	1	0.16
Business residence	5	0.78	8	1.25
Life services	0	0.00	1	0.16
Sports and leisure services	0	0.00	0	0.00
Healthcare services	0	0.00	0	0.00
Government agencies and social organisations	0	0.00	1	0.16
Accommodation services	8	1.25	4	0.62
Total	18	2.80	29	4.50

5.4 Spatial Inequity of Public Service Facilities in Macao and Hengqin

The Gini coefficient was used initially to reveal the spatial inequality. The Gini coefficient ranges from 0 (indicating perfect equity) to 1 (indicating perfect inequality). This study applies the Gini coefficient to measure the service capacity of each facility in Macao and Hengqin, which reflects the disparity in accessibility distribution. It thus provides a useful guide for decision making (Eq. 5.3) (Cheng et al., 2020).

$$G = \frac{1}{2N^2 \bar{A}} \sum_{m=1}^{N} \sum_{n=1}^{N} |A_m - A_n|, \qquad (5.3)$$

where G is the regional Gini coefficient, A_m and A_n correspond to the accessibility level of m and n zones; $\bar{A}$ is the average accessibility level of all zones in zone i; n is the total number of zones.

It is evident from Fig. 5.13 and Table 5.13 that the accessibility Gini coefficient in Macao is higher than that in Hengqin, indicating that the spatial variation of accessibility in Macao is more extensive and less equitable. The Areia Preta e Iao Hon in Macao, the population density is the highest, and the demand is higher. Therefore,

5.4 Spatial Inequity of Public Service Facilities in Macao and Hengqin

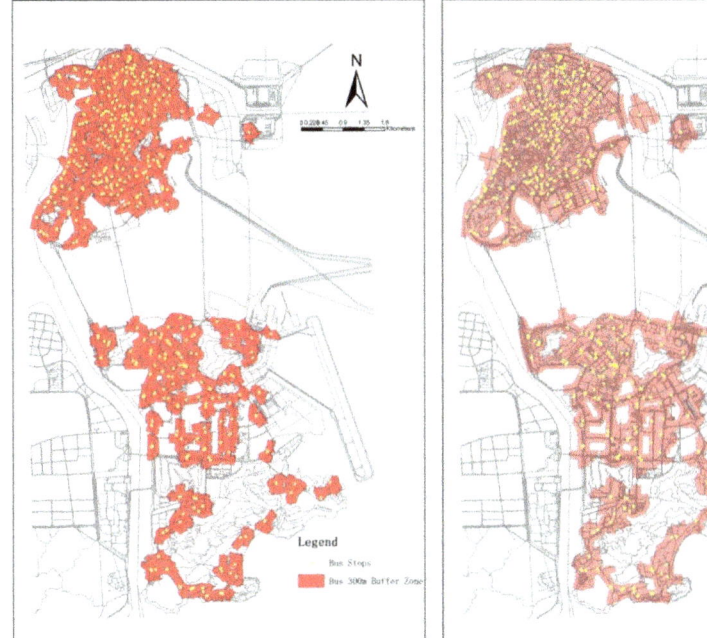

Fig. 5.4 Bus stations with 300- and 500-m service areas in Macao

Table 5.5 Bus station coverage in Macao

District Name	Population Coverage (%)		Area Coverage (%)	
	300 m	500 m	300 m	500 m
UOPG Norte–1	11.3	12.3	3.1	3.2
UOPG Norte–2	14.1	15.4	3.1	3.2
UOPG Este–1	12.8	14.2	3.8	4
UOPG Este–2	1.4	6.5	0.6	1
UOPG Central–1	15.6	15.6	3.5	3.5
UOPG Central–2	7.7	8	3.2	3.2
UOPG Central–3	6.5	7.8	2.9	3.3
UOPG Zona do Porto Exterior–1	0.2	0.3	1.5	1.7
UOPG Zona do Porto Exterior–2	1.7	2.1	6.8	7.1
UOPG Norte da Taipa–2	0.2	0.2	2.2	3.3
UOPG Taipa Central–1	0.7	0.7	2.9	3.4
UOPG Taipa Central–2	8.2	11.1	2	2.2
UOPG Pac On	1	2.1	7.2	7.5
UOPG Cotai	1	1.2	9.6	12.7
Total	82.4	97.5	52.4	59.3

Fig. 5.5 Bus stations with 300- and 500-m service areas in Hengqin

Table 5.6 Bus station coverage in Hengqin

Buffer area (m)	Covered population		Covered area	
	Quantity	Percentage	Quantity	Percentage
300	17,691	20.64	6.75	6.3
500	23,670	27.62	11.89	11.2

5.4 Spatial Inequity of Public Service Facilities in Macao and Hengqin

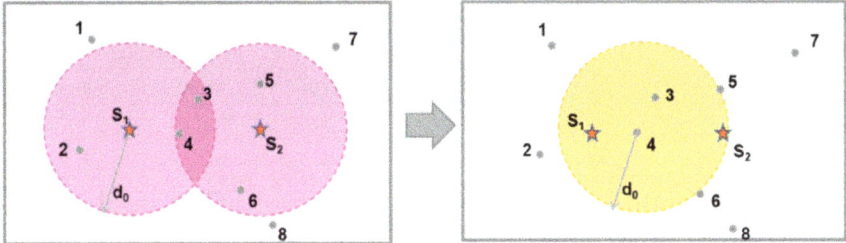

Fig. 5.6 2SFCA method

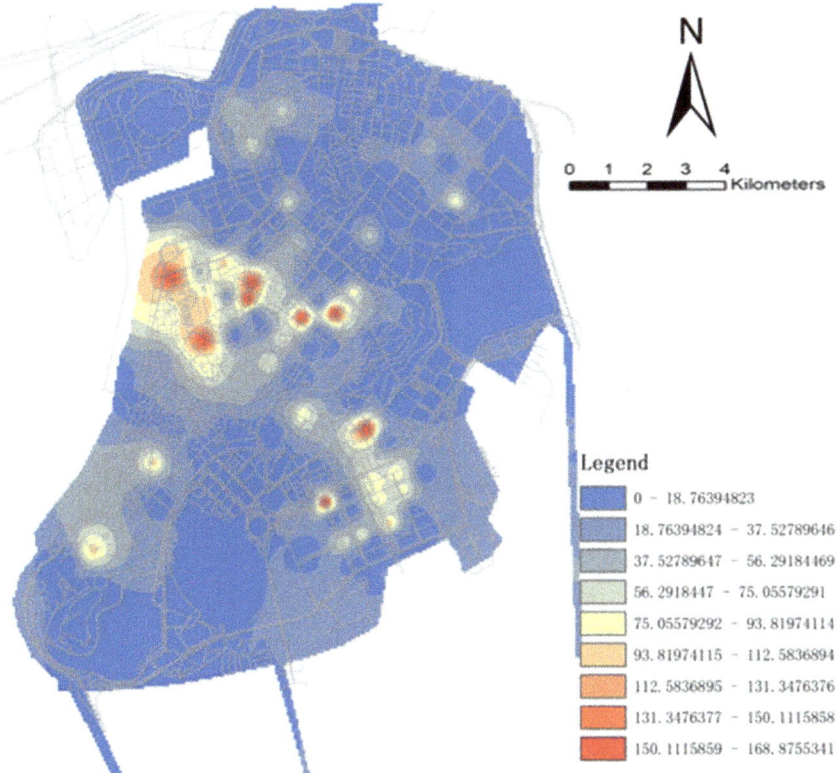

Fig. 5.7 Accessibility to schools in Macao

despite a large number of POI points, the accessibility is still low. Similarly, Hengqin is currently sparsely populated. Thus, the limited supply points can provide relatively good accessibility. In addition, Hengqin is still in the construction and development stage, and the settlement selected for this study is the built communities, which may be one of the reasons for the lower accessibility.

Table 5.7 Accessibility to schools in Macao

District name	Population coverage (%)		Area coverage (%)	
	300 m	500 m	300 m	500 m
UOPG Norte–1	11.3	12.3	3.1	3.2
UOPG Norte–2	14.1	15.4	3.1	3.2
UOPG Este–1	12.8	14.2	3.8	4
UOPG Este–2	1.4	6.5	0.6	1
UOPG Central–1	15.6	15.6	3.5	3.5
UOPG Central–2	7.7	8	3.2	3.2
UOPG Central–3	6.5	7.8	2.9	3.3
UOPG Zona do Porto Exterior–1	0.2	0.3	1.5	1.7
UOPG Zona do Porto Exterior–2	1.7	2.1	6.8	7.1
UOPG Norte da Taipa–2	0.2	0.2	2.2	3.3
UOPG Taipa Central–1	0.7	0.7	2.9	3.4
UOPG Taipa Central–2	8.2	11.1	2	2.2
UOPG Pac On	1	2.1	7.2	7.5
UOPG Cotai	1	1.2	9.6	12.7
Total	82.4	97.5	52.4	59.3

This chapter analyses the accessibility of public transport and public service facilities in Macao and Hengqin. The conclusion is that Macao's public transport system covers over half population and area percentage, whilst Hengqin's public transport radiation capacity is limited. It is suggested to strengthen the traffic connections, the coordination with new communities and the balance between work and living in the time dimension to better serve the daily life of residents.

5.4 Spatial Inequity of Public Service Facilities in Macao and Hengqin

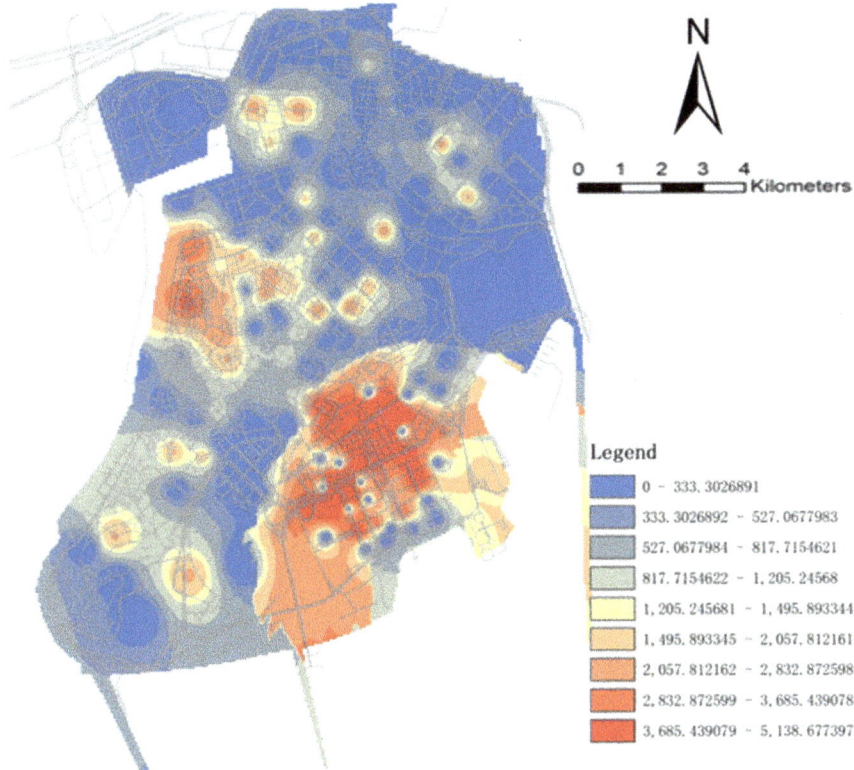

Fig. 5.8 Accessibility to bus stations in Macao

Table 5.8 Accessibility to bus stations in Macao

District Name	Area (m²)	Population		Accessibility
		Quantity	Percentage (%)	
UOPG Zona do Porto Exterior–2	7578.42	99	0.01	34,948.08
UOPG Zona do Porto Exterior–1	54,593.25	716	0.10	34,948.08
UOPG Zona do Porto Exterior–2	22,050.24	289	0.04	6764.43
UOPG Zona do Porto Exterior–2	14,250.89	187	0.03	4607.42
UOPG Central–2	124,242.41	8153	1.20	4292.63
UOPG Central–2	81,712.88	1072	0.16	4292.63
UOPG Zona do Porto Exterior–2	10,052.37	132	0.02	3762.64
UOPG Norte–1	46,245.82	4856	0.71	3664.92
UOPG Zona do Porto Exterior–2	29,703.97	390	0.06	3664.92
UOPG Zona do Porto Exterior–2	27,317.7	359	0.05	3129.52

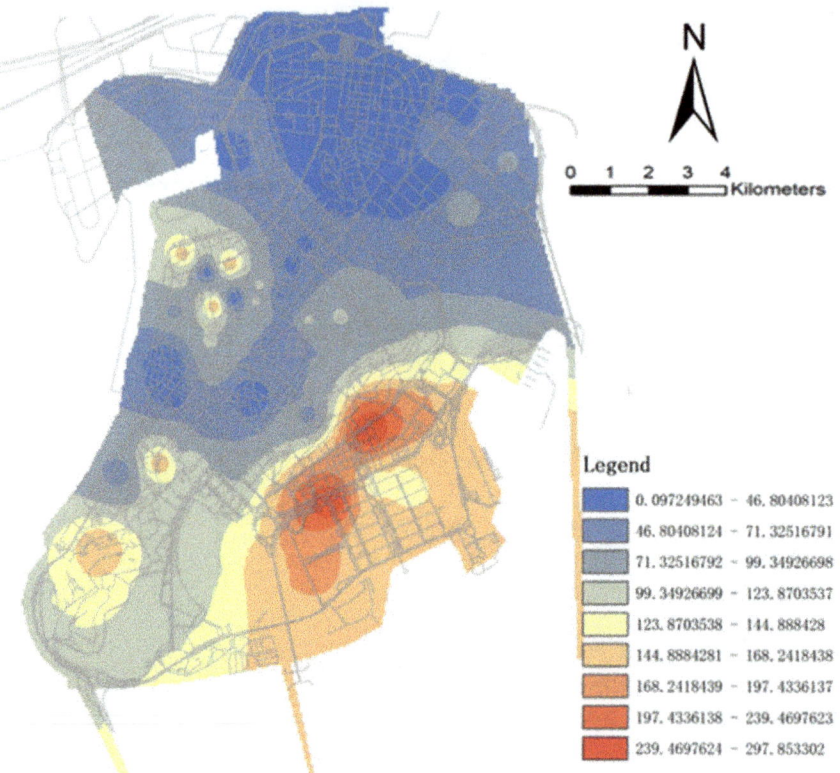

Fig. 5.9 Accessibility to other public service facilities in Macao

Table 5.9 Accessibility to other public service facilities in Macao

District name	Area (m^2)	Population		Accessibility
		Quantity	Percentage (%)	
UOPG Zona do Porto Exterior–2	27,317.7	359	0.05	147.97
UOPG Zona do Porto Exterior–2	29,364.17	385	0.06	145
UOPG Central–3	15,432.75	203	0.03	117.94
UOPG Central–3	170,917.92	11,216	1.60	59.71
UOPG Central–3	64,982.36	4264	0.60	20.52
UOPG Central–1	50,571.45	4646	0.70	47.4
UOPG Central–2	112,152.63	7360	1.00	90.48
UOPG Central–1	13,970.77	1284	0.20	59.71
UOPG Central–1	34,229.23	3145	0.50	164.77
UOPG Central–1	21,179.94	1946	0.30	104.72

5.4 Spatial Inequity of Public Service Facilities in Macao and Hengqin

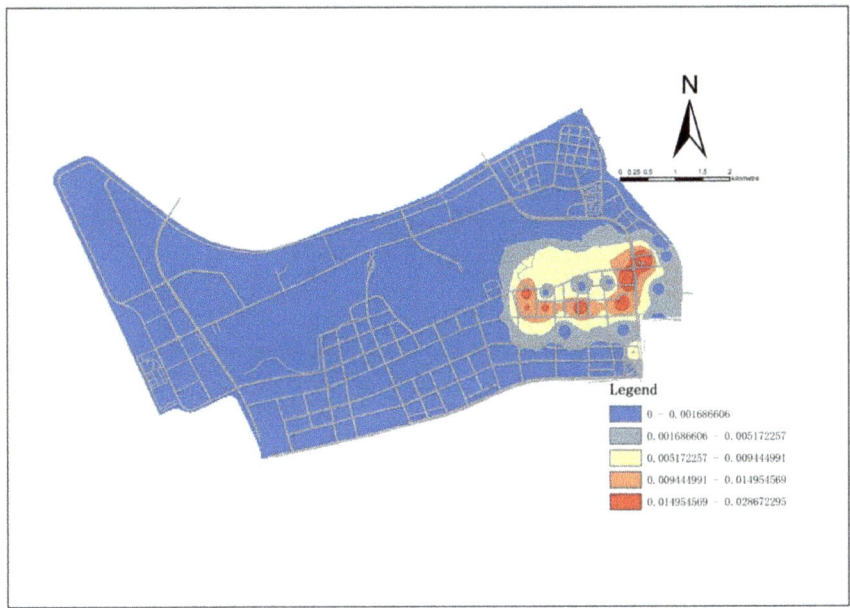

Fig. 5.10 Accessibility to schools in Hengqin

Table 5.10 Accessibility to school by residential area in Hengqin

Name of residence	Area (m²)	Population		Accessibility
		Quantity	Percentage (%)	
K3 Lai Chi Wan South	1,150,000	9822	11.50	20.04
Hua Fa Shou Fu West	1,290,000	4242	5.00	18.91
K2 Lai Chi Wan North	1,150,000	9822	11.50	17.66
Zhong Hai Ming Zuan North	960,000	4014	5.00	15.94
Hua Rong Qin Hai Bay West	1,200,000	5448	6.30	15.73
Long Guang Jiu long Seal West	500,000	3288	3.80	14.31
Yinxin Garden West	250,000	1098	1.30	11.93
Hengqin Xinjia South	480,000	3627	4.20	10.98

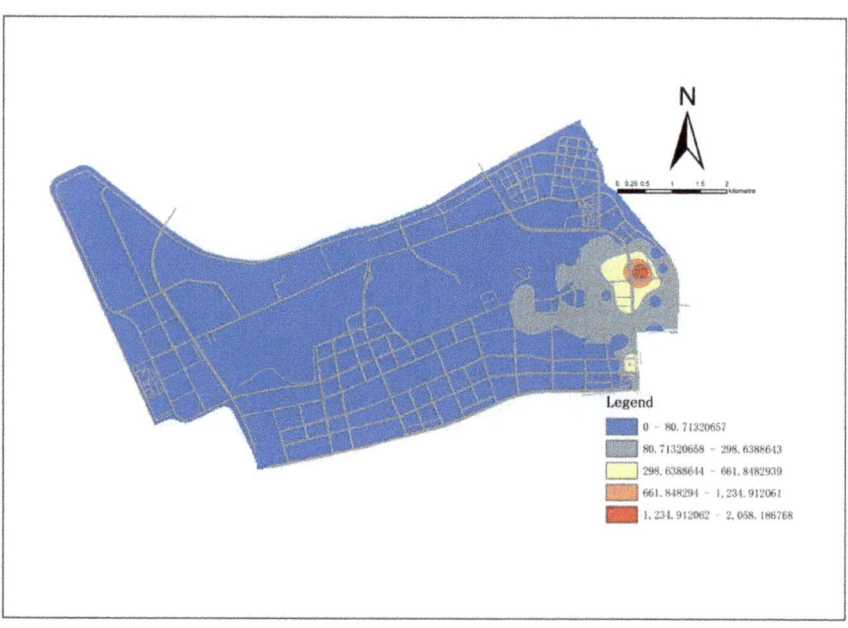

Fig. 5.11 Accessibility to bus stations in Hengqin

Table 5.11 Accessibility to bus stations in Hengqin

Name of residence	Area (m²)	Population		Accessibility
		Quantity	Percentage (%)	
Hengqin Xinjia South	480,000	3627	4.20	617.67
K3 Lai Chi Wan South	1,150,000	9822	11.50	123.45
Hua Fa Shou Fu West	1,290,000	4242	5.00	118.3
Zhong Hai Ming Zuan North	960,000	4014	4.70	241.6
Hua Rong Qin Hai Bay West	1,200,000	5448	6.40	293.52
K2 Lai Chi Wan North	1,150,000	9822	11.40	201.2
Long Guang Jiu long Seal West	500,000	3288	3.80	581.97
Yinxin Garden West	250,000	1098	1.30	2069.19

5.4 Spatial Inequity of Public Service Facilities in Macao and Hengqin

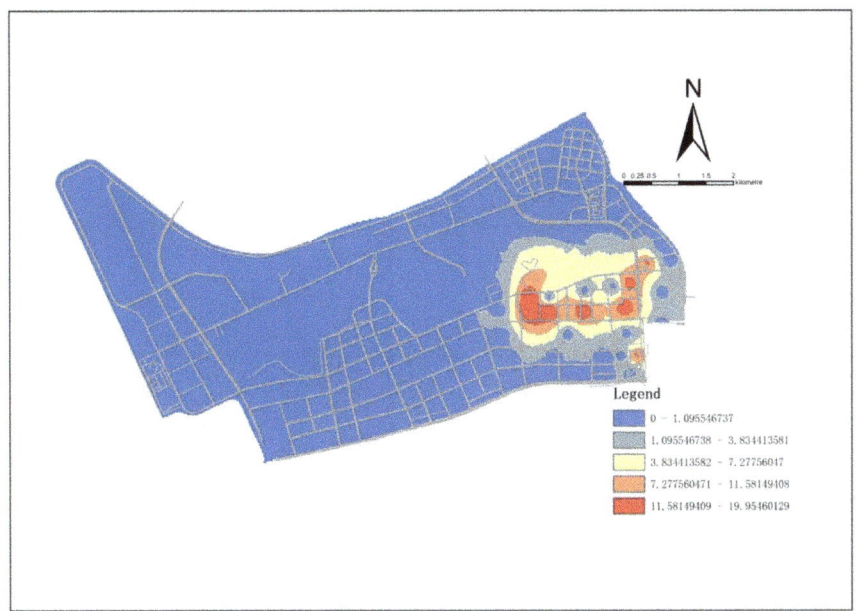

Fig. 5.12 Accessibility to other public service facilities in Hengqin

Table 5.12 Accessibility to other public service facilities in Hengqin

Name of residence	Area (m^2)	Population		Accessibility
		Quantity	Percentage (%)	
Yinxin Garden West	250,000	1098	1.30	0.02881
Long Guang Jiu long Seal West	500,000	3288	3.80	0.02342
Zhong Hai Ming Zuan North	960,000	4014	4.70	0.01957
Hua Rong Qin Hai Bay West	1,200,000	5448	6.30	0.01918
K2 Lai Chi Wan North	1,150,000	9822	11.40	0.01799
Hua Fa Shou Fu West	1,290,000	4242	5.00	0.01666
K3 Lai Chi Wan South	1,150,000	9822	11.50	0.01558
Hengqin Xinjia South	480,000	3627	4.20	0.00921

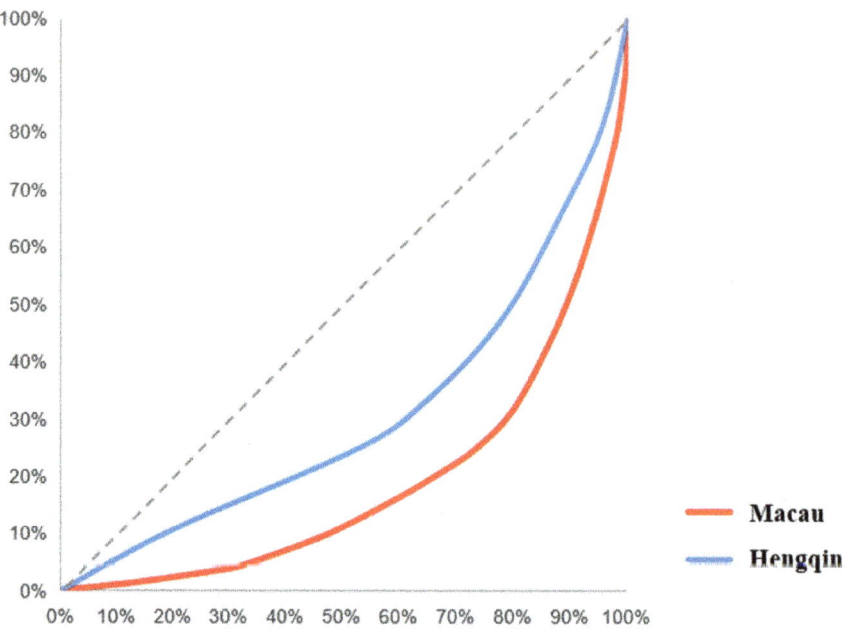

Fig. 5.13 Gini coefficients of accessibility in Macao and Hengqin

Table 5.13 Spatial concentrations of Macao and Hengqin

Area	Gini coefficient
Macao	0.62
Hengqin	0.53

References

Cheng, L., Yang, M., De, V. J., & Witlox, F. (2020). Examining geographical accessibility to multi-tier hospital care services for the elderly: A focus on spatial equity. *Journal of Transport & Health, 19*(2020), 100926. https://doi.org/10.1016/j.jth.2020.100926

Xing, L., Liu, Y., Wang, B., Wang, Y., & Liu, H. (2020). An environmental justice study on spatial access to parks for youth by using an improved 2SFCA method in Wuhan, China. *Cities, 96*(2020), 102405. https://doi.org/10.1016/j.cities.2019.102405

Zhang, Q. J., Guan, K. X., & Wang, H. X. (2022). Evaluation of spatial pattern characteristics and convenience of tourism resources in Zhengzhou. *Advance Online Publication.* https://kns.cnki.net/kcms/detail/31.1626.P.20220224.1710.002.html

Open Access This chapter is licensed under the terms of the Creative Commons Attribution 4.0 International License (http://creativecommons.org/licenses/by/4.0/), which permits use, sharing, adaptation, distribution and reproduction in any medium or format, as long as you give appropriate credit to the original author(s) and the source, provide a link to the Creative Commons license and indicate if changes were made.

The images or other third party material in this chapter are included in the chapter's Creative Commons license, unless indicated otherwise in a credit line to the material. If material is not included in the chapter's Creative Commons license and your intended use is not permitted by statutory regulation or exceeds the permitted use, you will need to obtain permission directly from the copyright holder.

Chapter 6
Ecosystem Services Analysis and Integration in Hengqin and Macao

Abstract With the rapid development of the city and economy, a series of environmental problems have emerged in Macao, such as the lack of freshwater resources, disastrous urban flooding caused by typhoons and heavy rains, intensification of the urban heat island effect, increased carbon dioxide emissions and pressure on recreational space. Hengqin is committed to promote Macao economy's sustainable development, facilitate urban integration of Hengqin–Macao and improve locals' welfare. In the urbanisation process, Hengqin is based on long-term scientific development, ensuring the coordination of development and ecological protection and striving to achieve the goal of 'ecological island'. As Macao is densely populated, there are scarce ecological resources with desperate needs of various ecosystem services. It is of great necessity to study the complementary interface between ecosystem services of Macao and Hengqin.

Keyword Ecosystem services · Integration and synergy · Hengqin · Macao

6.1 Macao and Hengqin Ecological Services Assessment

Ecosystem services are the benefits people obtain from ecosystems. These include provisioning, regulating, cultural and supporting services (MEA, 2003). Urbanisation poses challenges to providing and maintaining urban ecosystem services (Tzoulas et al., 2007). Current and projected climate change puts additional stress on urban environments by increasing heat waves, droughts, floods and water supply problems (IPCC, 2007). These issues present new challenges for urban planners to integrate concepts, such as ecosystem function, resilience, sustainability and biodiversity, into urban governance agendas and policies (FAO, 2011; Hansen et al., 2015). Humans' reliance on ecosystem services stems from ecosystems' supply capacity and social demand for these services. With the impact of urbanisation on ecosystem processes and social order, the supply and demand system of ecosystem services is becoming out of balance, resulting in various problems. By assessing the value of ecosystem services, this section has a theoretical basis for a more rational future allocation of urban green infrastructure resources and the sustainable management of ecosystem services in both Macao and Hengqin.

Table 6.1 Index system for assessing the value of terrestrial ecosystem services in Macao and Hengqin

Ecosystem level 1 services	Ecosystem level 2 services
Provisioning services	Soil formation
	Protecting species diversity
Regulating services	Climate regulation
	Purify the environment
	Noise reduction
	Gas Regulation
Cultural services	Recreational

Hengqin is committed to promote Macao economy's sustainable development, facilitate urban integration of Hengqin–Macao and improve locals' welfare. Hengqin has developed from a small island with banana forests and few farms to a modern and prosperous city. One of the goals of constructing Guangdong–Macao In-Depth Cooperation Zone in Hengqin is to create an 'ecological island' that is pleasant to live and work in. In this study, for seven terrestrial ecosystem services, a system of assessment indicators and accounting methods were selected and constructed to assess the amount of value of each terrestrial ecosystem service system (Table 6.1). The data source used in this study section are the 2020 Google Earth Engine 10-m resolution land use classification, which includes urban green infrastructures, built-up lands and unbuilt-up lands.

6.1.1 Soil Formation

Soil conservation is a conservation measure to prevent soil erosion, reduced soil productivity, acidification, salinization or other types of soil contamination caused by overuse of the soil. The soil conservation function of urban green space is divided into two aspects: soil holding and land fertility maintenance. The root system of various plants can effectively keep the soil and improve the soil's structure, porosity and permeability, which then can absorb more water and reduce surface runoff. By means of the transpiration of leaves, plants can adjust the temperature and humidity of the surrounding environment. Therefore, people's degree of comfort from the environment would be improved. At the same time, evaporation on the water surface bears away considerable heat, and water surface evaporation can also effectively regulate the local microclimate.

The potential and actual soil erosion difference method is used to calculate the functional amount of soil services sequestered by urban green space (Eq. 6.1). Its monetary value amount is assessed using the Shadow project method with the following equation (Eq. 6.2).

$$E_{\text{solid support}} = S_{\text{solid support}} \cdot T/Z, \qquad (6.1)$$

$$S_{\text{solid support}} = (A_f + k \cdot A_g)(Q_2 - Q_1), \quad (6.2)$$

where $E_{\text{solid support}}$ represents the value of soil service function of urban green space ecosystem; $S_{\text{solid support}}$ represents the amount of soil service function of urban green space ecosystem; T is the cost of earthwork per unit volume and is estimated according to the market experience value; Z is the soil capacity of urban green space in the study area. $S_{\text{solid support}}$ is the area of woodland in the study area; A_f is the area of grassland in the study area; k is the correction coefficient of grassland area; Q_1 and Q_2 are the soil erosion modulus with vegetation greening and without vegetation greening.

6.1.2 Protecting Species Diversity

Biodiversity is the collective term for all life forms on Earth, including species diversity, genetic diversity and ecosystem diversity. Amongst them, species diversity is the core, which reflects both the complex relationship between the richness of biological resources and surrounding environments (Wei et al., 2014). Species diversity embodies the organic link between species and the ecological environment in an ecosystem. Urban biodiversity is the basis for maintaining the balance of ecosystems. Rapid urban development and human activities have negatively affected the biodiversity of urban ecosystems, affecting the stability of urban ecological environments. As living places for various organisms, green spaces in cities can effectively protect species diversity.

The annual opportunity cost of species lost from an ecosystem is used to calculate the value of ecosystem services for the conservation of species diversity, which is calculated as follows.

$$E_g = (A_f + k \cdot A_g) J, \quad (6.3)$$

where E_g is the total value of the ecosystem conservation species diversity function in the study area; A_f is the area of woodland in the study area; A_g is the area of grassland in the study area; k is the correction factor of grassland area; J is the opportunity cost of species loss per unit area of urban green space.

6.1.3 Climate Regulation

Ecosystems influence temperature and precipitation at the local scale and regulate climate at the global scale by absorbing or emitting greenhouse gases, providing a climate suitable for human survival (Zhang et al., 2010). Plant transpiration and water surface evaporation are the main manifestations of the ecosystem's regulatory climate services. With the transpiration of leaves, plants can regulate the temperature and humidity of the surrounding environment, thus enhancing people's environmental comfort. At the same time, water surface evaporation can carry away a large amount of heat; thus, water surface evaporation can also effectively regulate the local microclimate and achieve the purpose of cooling the environment.

According to the current market, the value of plant transpiration is assessed by the 'avoid-cost' of electricity consumed by air conditioning and cooling to reduce the temperature. The value of the microclimate regulating function of the ecosystem is calculated as follows:

$$E_c = E_v + E_w, \tag{6.4}$$

$$E_v = (A_f + k \cdot A_a) \cdot H_a \cdot d \cdot \rho \cdot \alpha \cdot P_e, \tag{6.5}$$

$$E_w = W_a \cdot E_p \cdot \beta \cdot \rho \cdot \alpha \cdot P_e, \tag{6.6}$$

where E_c is the value of the ecosystem regulating microclimate function; E_V is the value of plant transpiration; E_W is the value of water surface evaporation; A_f is the area of woodland; A_a is the area of grassland (hm^2); k is the correction coefficient of grassland area; H_a is the heat absorbed per unit area of green space per day in summer, taken as 4.59 × 105 kJ/hm^2; d is the number of cooling days in summer, taken as 60 days; ρ is the constant 1 kWh/3600 kJ; a is the air conditioning efficiency ratio, taken as a conservative value of 3.0; P_e is the price of electricity; E_w is the area of the watershed (hm); E_p is the local annual evaporation (mm); β is the heat absorbed by evaporating unit volume of water 2.3 × 103 kJ/m^3.

6.1.4 Environment Purification

Green plants in an ecosystem use their metabolism to break down harmful chemicals, thereby reducing the concentration, quantity and toxicity of pollutants in the environment. The environmental purification function of ecosystems is reflected in two major areas: atmospheric purification and water purification. Plants on land and in waters can absorb pollutants in the air or water through the stomata on their leaves and realise the harmless conversion and discharge of pollutants via the assimilation and transfer functions of plants. Meanwhile, dense vegetation can slow down the speed at which pollutants, such as soot and dust, drift with the wind. Plants increase the humidity around the vegetation and on the leaf surface due to their transpiration, making it easier for dust pollution to be absorbed and descended. In this study, the area-absorptive capacity method was used to calculate the mass of atmospheric substances purified by the ecosystem in the study area, which was calculated as follows:

6.1 Macao and Hengqin Ecological Services Assessment

$$P_a = (A_f + A_g)(Q_{So2} + Q_{No2}), \quad (6.7)$$

where P_as is the total amount of atmospheric substances purified by forest and grassland in the study area (t), Q_{So2} is the amount of SO_2 absorbed per unit area of green space (kg/hm²), Q_{No2} is the amount of NO_2 absorbed per unit area of green space (kg/hm²).

6.1.5 Noise Reduction

Population gathering, urban traffic and various business stores generate a large amount of noise which significantly endangers people's physical and mental health if it exceeds a certain limit. Urban greenery has become an effective and economical control measure to urban noise. The approach to assess the amount of ecological services of urban green space for urban noise is as follows:

$$E_v = L_v \cdot P_v, \quad (6.8)$$

$$L_v = A_{\text{woodland}}/(0.04 * 100), \quad (6.9)$$

where E_V is the total value of the noise reduction ecosystem service in the study area, L_v is the woodland area of the urban green space converted into the length of the noise wall, P_v is the cost of noise control and is calculated according to the construction cost of each kilometre-long noise wall. A_{woodland} is the woodland area of the study area.

6.1.6 Climate Regulation

The importance of controlling carbon emissions has been raised to a new level by 'peak carbon dioxide emissions' and 'carbon neutrality' mentioned in the government work report in 2021. Using the carbon tax method and the Shadow project method to calculate the functional value of climate regulation services in Macao and Hengqin, the total value of climate regulation ecosystem services is calculated as follows:

$$E_{\text{cla}} = W_{\text{Fixed carbon}} \cdot P_{\text{Fixed carbon}} + W_{\text{release oxygen}} \cdot P_{\text{release oxygen}}, \quad (6.10)$$

where E_{cla} is the value of the climate regulation ecosystem service in the study area, $W_{\text{fixed carbon}}$ is the price per unit weight of carbon dioxide emissions levied on the tax rate, $P_{\text{release oxygen}}$ is the price per unit weight of oxygen production.

6.1.7 Cultural Service

The cultural value of ecosystem services refers to the indirect benefits of human beings in terms of aesthetic interest, cultural science, recreation and leisure and spiritual emotion. As an indispensable outdoor leisure space for urban residents, urban green space can provide people with enjoyment and cultural value and benefit the mental health condition of residents. Besides, urban green space forms a unique urban natural landscape and regional landscape, highlights cultural characteristics of the region and provides the fundamental resource for local tourism development.

Based on the estimated unit area value equivalent to Chinese ecosystems (Xie et al., 2003), the cultural ecosystem service value of the study area is estimated as follows:

$$EPV = A_k \cdot VC_k, \tag{6.11}$$

where EPV is the total value of ecosystem cultural service in the study area, A_k is the total area of the ten land use types of category k and VC_k is the value coefficient of cultural ecosystem service per unit area of category k land use type.

6.2 Analysis of the Value of Terrestrial Ecological Services in Macao and Hengqin

A comparison of terrestrial ecosystem services between Macao and Hengqin is shown in Table 6.2 and Fig. 6.1. The ecosystem value of Hengqin is nearly three times higher than that of Macao. The areas with high ecological value in Macao are Parque de Merendas da Barragem de Ká-Hó, Parque Natural da Taipa Grande and Flora Garden. The areas with high ecological value in Hengqin are in the area of Da Hengqin Island and Xiao Hengqin Island. The value per square kilometre is RMB 190.77 million per year in Macao and RMB 379.90 million per year in Hengqin. The value per square kilometre in Hengqin is higher than that in Macao. By deploying the Hengqin Master Development Plan, Hengqin has delineated spatial control zones in the No Build Zone, Restricted Build Zone, Suitable Build Zone, which strictly protects the pristine ecological environment of Hengqin.

6.3 Vision for the Future of Macao and Hengqin

'Building a low-carbon Macao and creating a green life' is the vision of Macao's ecological and environmental protection, which guides the sustainable development of Macao and provides strong support for Macao to accelerate its integration into regional and national development. The 'Joint Response to Climate Change: Building

6.3 Vision for the Future of Macao and Hengqin

Table 6.2 Value composition of terrestrial ecosystem services in Macao and Hengqin in 2020 (RMB million)

Ecosystem level 1 services	Ecosystem level 2 services	Macao	Hengqin
Provisioning services	Soil formation	100.4	4926.98
	Protecting species diversity	850.41	4465.35
Regulating services	Climate regulation	1505.47	11,984.59
	Purify the environment	322.38	10,342.36
	Noise reduction	694.13	
	Gas Regulation	2449.45	4245.93
Cultural services	Recreational	487.91	4330.89
Total		6410.00	40,270.00
Value per km^2		19,077.00	37,990.00

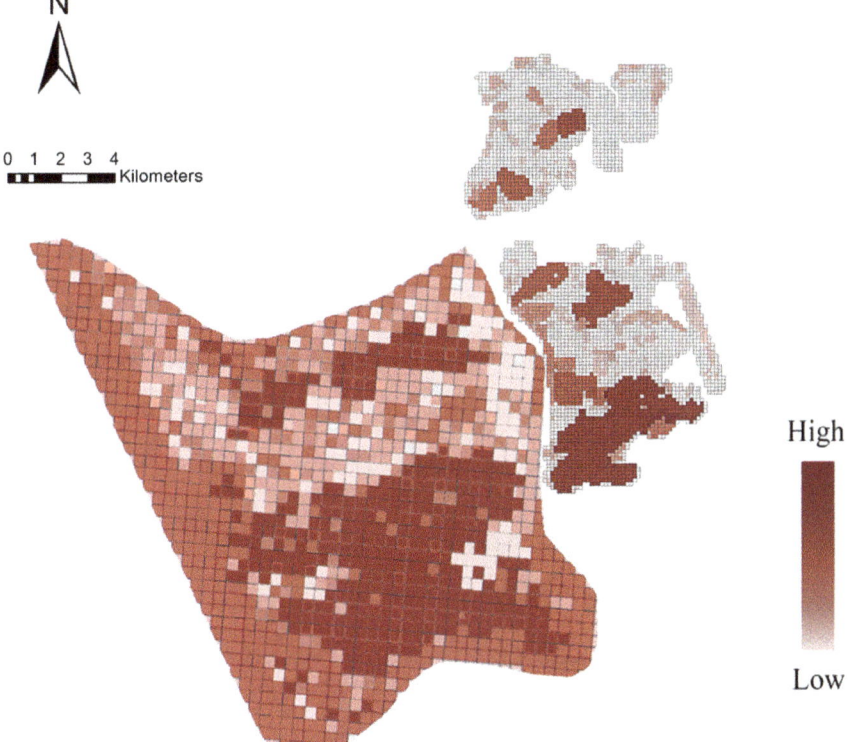

Fig. 6.1 Comparison of the value of Macao and Hengqin land ecological service system

a Green and Low-Carbon Macao', 'Strengthening Environmental Pollution Control: Building a Liveable and Tourable City' and 'Strengthening Ecological Environmental Protection: Enhancing the Quality of Life in Macao' are the critical work of environmental protection in Macao. At present, Hengqin has developed from a wild island to a modern city with an excellent ecological environment, infrastructure and diversified cultural life. The ecological value of Hengqin is high per unit area, and a comparison of ecological values within construction lands needs further analysis and research.

References

FAO. (2011). State of Europe's forests 2011. Status and trends in sustainable forest management in Europe. In *Ministerial conference on the protection of forests in Europe*. Retrieved June 30, 2022, from https://unece.org/fileadmin/DAM/publications/timber/Forest_Europe_report_2011_web.pdf

Hansen, R., Frantzeskaki, N., McPhearson, T., Rall, E., Kabisch, N., Kaczorowska, A., Kain, J.-H., Artmann, M., & Pauleit, S. (2015). The uptake of the ecosystem services concept in planning discourses of European and American cities. *Ecosystm Services, 2015*(12), 228–246. https://doi.org/10.1016/j.ecoser.2014.11.013

IPCC. (2007). *Climate change 2007-impacts, adaptation and vulnerability: Working group II contribution to the fourth assessment report of The IPCC*. Cambridge University Press 2007. Retrieved June 30, 2022, from https://www.ipcc.ch/report/ar4/wg2/

Millennium ecosystem assessment, M. E. A. (2005). *Ecosystems and human well-being*, Vol. 5 (pp. 563–563). Island press.

Tzoulas, K., Korpela, K., Venn, S., Yli-Pelkonen, V., Kaźmierczak, A., Niemela, J., & James, P. (2007). Promoting ecosystem and human health in urban areas using Green Infrastructure: A literature review. *Landscape and Urban Planning, 81*(3), 167–178. https://doi.org/10.1016/j.landurbplan.2007.02.001

Wei, W. F., Nie, G. Y., Miao, X. H., Lu, H., & Hu, B. Y. (2014). Progress in the study of biodiversity loss mechanisms. *Scientific Bulletin, 59*(6), 430–437.

Xie, D. G., Lu, X. C., Len, F. Y., Zheng, D., & Li, C. S. (2003). Valuation of ecological assets on the Qinghai-Tibet Plateau. *Journal of Natural Resource, 02*, 189–196.

Zhang, B., Xie, G. D., Xiao, Y., & Lun, F. (2010). Classification of ecosystem services based on human demand. *China Population, Resources and Environment, 20*(6), 64–67.

Open Access This chapter is licensed under the terms of the Creative Commons Attribution 4.0 International License (http://creativecommons.org/licenses/by/4.0/), which permits use, sharing, adaptation, distribution and reproduction in any medium or format, as long as you give appropriate credit to the original author(s) and the source, provide a link to the Creative Commons license and indicate if changes were made.

The images or other third party material in this chapter are included in the chapter's Creative Commons license, unless indicated otherwise in a credit line to the material. If material is not included in the chapter's Creative Commons license and your intended use is not permitted by statutory regulation or exceeds the permitted use, you will need to obtain permission directly from the copyright holder.

GPSR Compliance

The European Union's (EU) General Product Safety Regulation (GPSR) is a set of rules that requires consumer products to be safe and our obligations to ensure this.

If you have any concerns about our products, you can contact us on

ProductSafety@springernature.com

In case Publisher is established outside the EU, the EU authorized representative is:

Springer Nature Customer Service Center GmbH
Europaplatz 3
69115 Heidelberg, Germany